71st Northeastern Weed Science Society Annual Meeting 2017

Proceedings – Volume 71

Philadelphia, Pennsylvania, USA
3 – 6 January 2017

Editor:

Michael L. Flessner

ISBN: 978-1-5108-4928-0

Printed from e-media with permission by:

Curran Associates, Inc.
57 Morehouse Lane
Red Hook, NY 12571

CURRAN ASSOCIATES INC. proceedings.com

Some format issues inherent in the e-media version may also appear in this print version.

Printed by Curran Associates, Inc. (2017)

Northeastern Weed Science Society (NEWSS)
P.O. Box 25
Woodstown, NJ 08098
USA

Phone: 315-209-7580

northeasternweedscience@hotmail.com

Additional copies of this publication are available from:

Curran Associates, Inc.
57 Morehouse Lane
Red Hook, NY 12571 USA
Phone: 845-758-0400
Fax: 845-758-2633
Email: curran@proceedings.com
Web: www.proceedings.com

PROCEEDINGS - VOLUME 71

71st Annual Meeting of the Northeastern Weed Science Society

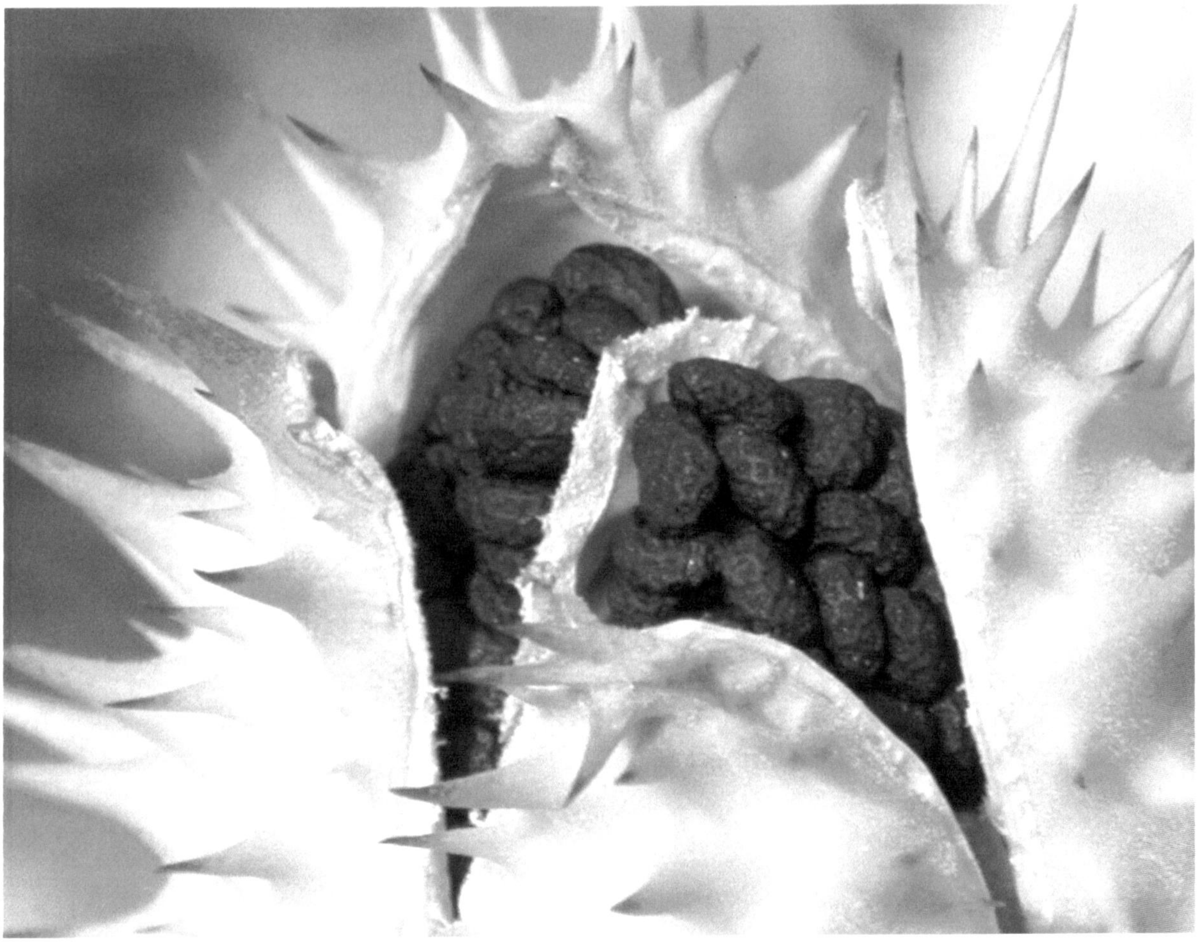

January 3-6, 2017 – Sheraton Society Hill - Philadelphia, PA

<u>Site of 2018 Joint Meeting*</u>

January 9-11, 2018

Hilton Philadelphia at Penn's Landing
201 South Columbus Boulevard
Philadelphia, Pennsylvania 19106

Participating Societies:
Northeastern Weed Science Society
American Society of Agronomy – Northeastern Regional Branch
American Society of Horticultural Science – Northeast Division

Proceedings

of the

Seventy First Annual Meeting

of the

Northeastern Weed Science Society

Michael L Flessner, Editor

NORTHEASTERN WEED SCIENCE SOCIETY

2017 SUSTAINING MEMBERS

Platinum

BASF The Chemical Company

syngenta

Gold

VALENT

FMC

Bayer CropScience

MONSANTO

Silver

Gowan USA The Go To Company

ohp Partners with solutions

UPI

GDM

Bronze

AMVAC — Crop Management Strategies

ACDS — TeeJet® Technologies

LABServices

Table of Contents

NEWSS EXECUTIVE COMMITTEE – OFFICERS

President	**Randy Prostak** University of Massachusetts rprostak@umext.umass.edu
President-elect	**Carroll Moseley** Syngenta Carroll.moseley@syngenta.com
Vice President	**Dan Kunkel** Rutgers University kunkel@aesop.rutgers.edu
Treasurer	**Kurt Vollmer** University of Delaware kvollmer@udel.edu
Past President	**Shawn Askew** Virginia Tech saskew@vt.edu

NEWSS EXECUTIVE COMMITTEE – MEMBERS

Membership Chair	**Sudeep Mathew** Syngenta sudeep.mathew@syngenta.com
Editor	**Michael Flessner** Virginia Tech flessner@vt.edu
Public Relations	**Katelyn Venner** BASF katelyn.venner@basf.com
Research and Education Coordinator	**Quintin Johnson** University of Delaware quintin@udel.edu
Sustaining Membership	**Wesley Everman** North Carolina State University wes_everman@ncsu.edu
CAST Representative	**Rob Richardson** North Carolina State University rob_richardson@ncsu.edu
Graduate Student Representative	**Theresa Reinhardt** North Carolina State University tareinha@ncsu.edu
WSSA Representative	**Mike Fidanza** The Pennsylvania State University maf100@psu.edu
Science Policy Director	**Lee Van Wychen** Weed Science Society of America lee.vanwychen@wssa.net

2017 NEWSS MEETING PROGRAM COMMITTEES

Program Committee	Chair:	Carroll Moseley, Syngenta Dan Kunkel, Rutgers University
Agronomy	Chair:	Sandeep Rana, North Carolina State University
Vegetation Management and Resortaion	Chair:	Lane Heimer, Maryland Dept. of Agriculture
Ornamentals	Chair:	Sali Barolli
Research Posters	Chair:	Michael Flessner, Virginia Tech
Turfgrass and Plant Growth Regulators	Chair:	Matthew Elmore, Rutgers University
Vegetables and Fruit	Chair: Chair-elect:	Thierry Besancon, Rutgers University Charlie Cahoon, Virginia Tech
Weed Biology and Ecology	Chair:	Caroline Marschner, Cornell University

2017 NEWSS COMMITTEES

Student Paper Contest	Chair:	Greg Armel, BASF
Student Poster Contest	Chair:	Darren Lycan, Syngenta
Photo Judging Contest	Chair:	Javier Vargas, University of Tennessee
Site Selection Committee	Chair: Chair-elect:	Carroll Moseley, Syngenta Randy Prostak, University of Massachusetts
Awards Committee	Chair:	Shawn Askew, Virginia Tech
Nomination Committee	Chair:	Shawn Askew, Virginia Tech
Resolutions Committee	Chair:	Burkhard Schulz, University of Maryland
Collegiate Weed Contest	Chair:	Carroll Moseley, Syngenta
Audit Committee	Chair:	Kurt Vollmer, University of Delaware
Archives Committee	Chair:	Dan Kunkel, Rutgers University
Weed Science Field Days	Chair:	Quintin Johnson, University of Delaware
Past Presidents Committee	Chair:	Shawn Askew, Virginia Tech
Herbicide Resistance	Chair:	Mark VanGessel, University of Delaware
Endowment Committee	Chair:	Randy Prostak, University of Massachusetts

2017 AWARDS, NEWSS ANNUAL MEETING

NEWSS Fellow Award - Hilary Sandler

Award of Merit - Robin Bellinder

Award of Merit - Greg Armel

Outstanding Researcher Award - Wes Everman

Outstanding Educator Award - Jim Brosnan

Robert D. Sweet Outstanding Ph.D. Student Award - Katelyn Venner

Robert D. Sweet Outstanding M.S. Student Award - Jeff Liebert

NEWSS CONTEST – WINNERS, 2017

STUDENT PAPER

1st place paper – Cody Dickinson

2nd place paper – Jeffrey Liebert

Honorable mention – Steven Haring

Honorable mention – Rebecca Champagne

STUDENT POSTER

1st place poster – Brandon Schrage

2nd place poster – John Sanders

PHOTO CONTEST

1st place photo – Antonio DiTommaso

2nd place photo – Connor Youngman

3rd place photo – Eugene Law

Honorable mention – Antonio DiTommaso

Honorable mention – Jennifer D'Appollonio

2017 NEWSS COLLEGIATE WEED SCIENCE CONTEST

Hosted by Virginia Tech, July 25th, 2017

(Photos courtesy of Shawn Askew, Jordan Craft and Katelyn Venner)

The 2016 NEWSS Annual Weed Contest was hosted at Virginia Tech on July 26, 2016. Seven graduate teams and seven undergraduate teams competed as well as several individuals for a total of 52 students from four universities across the region. The contest went smoothly and enjoyed hot but dry weather. Many students and coaches mentioned that the contest site (Kentland Farm) was beautiful, and all the events were well done and very challenging. The contest concluded with the awards ceremony at Lane Stadium, where participants got a chance to venture onto Worsham Field. In an extremely close contest, the University of Guelph Graduate Team 1 won the graduate student division, just edging out North Carolina State University Team 1. The University of Guelph Undergraduate Team 1 won the undergraduate division. Overall, it was a tremendous success. It could not have been possible without financial support and many volunteers.

Farmer Problems

<u>Farmer Problem #1. Tall Fescue in Juniper Beds</u>

Farmer: RJ Richtmyer

Judge: Cody Dickinson

<u>Farmer Problem #2. Lawn at Brick House</u>

Farmer: Katelyn Venner

Judge: Sandeep Rana

Farmer Problems...cont'd

Farmer Problem #3. Golf Rough at Airport

Farmer: Blake Moore

Judge: Todd Davis

Farmer Problem #4. Soybean Stand Loss

Farmer: Art Gover

Judge: Kelly Patches

Farmer Problems...cont'd

Farmer Problem #5. Jimsonweed in Cow Manure

Farmer: Melissa Bravo

Judge: Kelly Liberator

Farmer Problem #6. Soybean Burn Injury

Farmer: Larissa Smith

Judge: Becky Fletcher

Farmer Problems…cont'd

Farmer Problem #7. Farmer Needs a Loan…

Farmer: Lloyd Hipkins

Judge: Shawn Beam

Farmer Problem #8. Sweet Corn.

Farmer: Charlie Cahoon

Judge: Bob Pitman

Visiting Worsham Field

Lane Stadium, Virginia Tech, Blacksburg VA

The Vice President of NEWSS, Carroll Mosley, chats with students on Worsham Field.

NEWSS President, Shawn Askew, explains the on-field games to eager contest attendees.

Treasurer and VT Alum, Kurt Vollmer, channeled his inner football player at Worsham Field.

Awards

Undergraduate Team Awards

First Place Undergraduate

Guelph Undergraduate Team 1

(Matt Rundle, Alexander Sanders, Hanna Petrovsky and Kaylene Sangers)

Second Place Undergraduate

Guelph Undergraduate Team 4

(Michael Palermo, Aaron Beunen, Ryan Carlow and Peter Buys)

Third Place Undergraduate

Guelph Undergraduate Team 2

(Laura Buys, Caleb Niemeyer, Amelia Judge and Bret Hilker)

Awards...cont'd

Undergraduate Individual Awards

First Place Undergraduate

Alexander Sanders (Guelph)

Second Place Undergraduate

Matt Rundle (Guelph)

Third Place Undergraduate

Brendan Metzger (Guelph)

Awards...cont'd

Graduate Student Team Awards

First Place Graduate Team

Guelph Graduate Team 1

(Mike Schryver, Meghan Grguric, Taiga Cholette, and Brittany Hedges)

Second Place Graduate Team

North Carolina State University Grad 1

(Nick Basinger, Drake Copeland, Matt Bertucci, and Anthony Growe)

Third Place Graduate Team

Pennsylvania State University Grad 1

(Jess Bunchek, Rebecca Champagne, Mitch Hunter, and Heidi Myer)

Awards…cont'd

Graduate Individual Awards

First Place Graduate Student

Drake Copeland (North Carolina State University)

Second Place Graduate Student

Mitch Hunter (Pennsylvania State University)

Third Place Graduate Student

Brandon Schrage (North Carolina State University)

WHEAT VARIETY TOLERANCE TO METRIBUZIN. M.L. Flessner* and W.E. Thomason, Virginia Tech, Blacksburg, VA

ABSTRACT

Metribuzin is labeled in several Mid-Atlantic States for group 2 resistant common chickweed (*Stellaria media* (L.) Vill.) control in wheat (*Triticum aestivum* L.). Metribuzin application has resulted in wheat injury, which varies by variety. Research was conducted to assess wheat variety tolerance to metribuzin, as little information is available in this regard. In 2015-16, a factorial experiment evaluated the response of 39 wheat varieties to metribuzin (TriCor) at 0 or 0.28 kg ha^{-1} at two locations (Campbell and New Kent counties, Virginia), respectively. Metribuzin was applied at the 3 to 4 tiller growth stage. No differences in visible injury were observed when comparing treated to nontreated plots within each variety 7 to 21 days after treatment (DAT). A second factorial study was conducted in 2016-17 in Montgomery County, Virginia. The same 39 varieties were evaluated for response to metribuzin at 0, 0.28, and 0.56 kg ha^{-1} applied at the 2 to 4 leaf stage. Metribuzin at 0.56 kg ha^{-1} resulted in 10 to 15% visible chlorosis to three varieties (VA11W-313, USG 3523, VA11W-279) 7 to 21 DAT; 12 varieties responded with 5 to 10% chlorosis. All other varieties responded with <5% chlorosis 7 to 21 DAT. Metribuzin at 0.28 kg ha^{-1} resulted in 6% chlorosis to VA11W-279, 7 to 21 DAT; all other varieties responded with <4% chlorosis. A third experiment was conducted in 2015-16 in Montgomery County, Virginia to evaluate the response of Shirley, Jamestown, and SS5205 wheat varieties to metribuzin at 0, 0.07, 0.14, 0.28, and 0.56 kg ha^{-1} with and without granular urea applied concurrently at 56 kg N ha^{-1}. Treatments were applied at the 3 to 4 leaf stage. Results indicate these wheat varieties are tolerant to metribuzin $\leq$ 0.28 kg ha^{-1}, which is the highest labeled rate as <5% visible chlorosis was observed 9 to 24 DAT. Metribuzin at 0.56 kg ha^{-1} resulted in 5 to 16% visible injury across varieties 9 to 24 DAT. Yield reductions of 693 kg ha^{-1} without N and 1,668 kg ha^{-1} with N were observed in Shirley only, compared to the nontreated and N only treated checks, respectively. No yield reductions were observed in Jamestown or SS5205. Across experiments, these results indicate that wheat varieties tested are tolerant to metribuzin at $\leq$ 0.28 kg ha^{-1} when applied at between 2 to 4 leaf and 2 to 4 tiller growth stages. Future research should evaluate metribuzin applied with surfactants and at earlier growth stages.

FOLIAR HERBICIDE APPLICATION SAN BE EFFECTIVE IN LATE FALL. J.S. Caplan*, R.D. Whitehead, A.E. Gover, and G.C. Grabosky, Rutgers University, New Brunswick, NJ

ABSTRACT

Numerous shrub species are invading temperate deciduous forests, and many are aided in their growth by delaying leaf senescence until late fall. Focusing on *Lonicera maackii*, we carried out a series of studies to evaluate the possibility that species retaining large portions of their canopies can be controlled using foliar herbicide application far later than it is typically attempted. We first evaluated the potential for susceptibility in four species that have green leaves in early November. We measured physiological parameters related to gas exchange, and found that rates of all parameters (e.g., light-saturated photosynthesis) were comparable to those measured in the same or similar shrubs in summer. To determine directly if an invasive species with extremely delayed leaf senescence (*L. maackii*) was susceptible to control in mid-November, we evaluated the effectiveness of three foliar sprays: glyphosate and two concentrations of aminocyclopyrachlor plus metsulfuron methyl. We found that treatments killed (72%) or severely injured (14%) of plants, but identified no statistically-significant differences among herbicide types. Finally, because efforts to treat plants in late fall will be guided primarily by visual assessments of leaf color until further trials are performed, we characterized how changes in color correspond with declining photosynthetic rates. A single regression equation was appropriate for all species, indicating that assessments of yellowing can determine when leaves are no longer photosynthetic. Our study demonstrates that late-fall can be a viable period in which to treat weedy species that delay senescence strongly, like the many invasive *Lonicera* species in North America. Given that climate change will delay senescence further in many species, our study also serves as a call for further investigation into what promises to be an increasingly viable opportunity for weed control in deciduous forests.

WATER TEMPERATURE AND EXPOSURE TIME FOR KILLING WEED SEED ON RECYCLED PLASTIC CONTAINERS. J. Altland*, USDA-ARS, Wooster, OH

ABSTRACT

Seeds of many weeds, most notably creeping woodsorrel (*Oxalis corniculata*) and bittercress (*Cardamine flexuosa*), adhere to plastic containers and trays and are reintroduced into the production system when the containers and trays are reused. Some nurseries use hot water or steam for sterilizing reused containers and propagation trays. Initially, they adopted this technology as a means for eliminating pathogens, but soon noted vast improvements in weed control. Nursery operations are using hot water or steam at temperatures ranging from 60 to 90 °C, with exposure times from 15 minutes to 4 hours. While they have reported increased levels of weed control using this form of sterilization, the temperature and exposure times selected were based on best guesses. Therefore, the objective of this research was to determine the specific temperatures and exposure times necessary to kill creeping woodsorrel and bittercress seeds using hot water.

Initial experiments with creeping woodsorrel and bittercress were conducted separately. Glass test tubes were filled with ten seeds each, then placed into a digitally programmable hot water bath. The hot water bath was set at 60, 75, or 90 °C for 1, 5, 10, 30, or 60 minutes to determine creeping woodsorrel and bittercress tolerance. There were five replicate test tubes per treatment, including a group of five control test tubes that remained at room temperature. After heat treatment, the seeds from each test tube were transferred to a Petri dish containing an agar base made using 15 g L^{-1} granulated agar in a modified Hoagland solution (in mM: 7.5 N, 0.5 P, 3 K, 2.5 Ca, 1 Mg, 1 S, 0.071 Fe, 0.009 Mn, 0.0015 Cu, 0.0015 Zn, 0.045 B, 0.0001 Mo, 0.024 Cl, and 0.0002 Na). Petri dishes were placed in a growth chamber providing a 12-hour photoperiod and 18 °C night/22 °C day air temperature. After 2 weeks, weed germination in each Petri dish was tabulated.

Creeping woodsorrel treated with 60 °C water had a similar germination percentage as non-heated controls. Those heated at 75 °C still germinated, but at a lower percentage than the non-heated controls. None germinated when exposed at 90 °C for 5 minutes or longer. Within each temperature, exposure time did not affect creeping woodsorrel germination. Similarly, bittercress were not controlled with 60 °C water. Bittercress germination was reduced when treated with 75 °C water, and germination decreased with increasing exposure time. Like creeping woodsorrel, none of the bittercress germinated when treated with 90 °C water. For rapid exposure times of 5 minutes or less, high temperatures of at least 90 °C will be needed for effective control of creeping woodsorrel and bittercress. Lower temperatures of 75 to 85 °C might be effective with sufficiently long exposure time.

TOLERANCE OF CONTAINER-GROWN DOGWOOD AND VIRGINIA SWEETSPIRE TO SP-1770 HERBICIDE. J.S. Aulakh*, The Connecticut Agricultural Experiment Station, Windsor, CT

ABSTRACT

SP-1770 is an experimental herbicide under evaluation for weed control efficacy and ornamental plants safety. Container studies were conducted at the Valley Laboratory of the Connecticut Agricultural Experiment Station at Windsor, CT to evaluate tolerance of Dogwood (*Cornus spp* 'Arctic fire') and Virginia sweetspire (*Itea virginica* 'Henry garnet') to over-the-top applications of SP-1770. Experiments were conducted in a completely randomized design with 15 replicates. Treatments consisted of 3 rates of SP-1770 (9.6, 19.2, and 38.4 floz/ac) and a non-treated control. Two sequential applications of SP-1770 were made at approximately 7 wk interval. Data were recorded on initial and final plant height, canopy width, and periodic plant injury, discoloration, and stunting. Results indicated significant necrotic injury to dogwood at all the rates tested while Virginia sweetspire was most injured at 38.4 floz/ac. The highest rate of 38.4 floz/ac was the most injurious; injury was rated 6.0 and 8.7 in dogwood and 4.1 and 6.0 in virgina sweetspire at 28 days after first and second application, respectively. To Virginia sweetspire, SP-1770 rates up to 19.6 floz/ac were comparatively less injurious (<2.0). All rates of SP-1770 resulted in commercially unacceptable levels of discoloration in Dogwood and Virginia sweetspire with a range of 1.9 to 7.0. However, significant plant stunting occurred only at 38.4 floz/ac.

EFFECT OF PRE-PLANT NITROGEN (N) RATES ON WHEAT YIELD IN CORN/SORGHUM-WHEAT ROTATION. M.K. Bansal* and W.J. Everman, North Carolina State University, Raleigh, NC

ABSTRACT

Sorghum production has gained interest in recent years as regional grain demands increased which lead swine producer to offer a competitive sorghum grain price. Sorghum can be a good alternative for corn in rotation with wheat. Sorghum has ability to tolerate hot dry weather, a condition that can be challenging for corn in drought season. However, with the advantages, sorghum has some disadvantages as well when used in rotation. Grain sorghum is known to have negative impact on the following crop. Sorghum residue when incorporated in soil can make N immobilize making it less available to following wheat.

Experiments were conducted in 2013-14 at Rocky Mount, 2014-15 at Rocky Mount and Kinston (two locations), and 2015-16 at Rocky Mount and Kinston, North Carolina to evaluate the effect of different rates of pre-plant nitrogen (15, 30, 45, and 60 lbs per acre) applied to wheat following different hybrids either sorghum (DKS 53-67, P83P17) or corn (DKC 60-67) on wheat yield. There was no significant effect of pre-plant nitrogen on wheat yield in both 2013-14 and 2015-16. In 2013-14, there was significant effect of hybrids on wheat yield. Wheat yield was not significantly different when planted after either DKC 60-67 or DKS 53-67. Yield was significantly different when planted after DKC 60-67 and P83P17. In 2014-15 and 2015-16, there was no significant effect of different hybrids on wheat yield at all locations. Pre-plant nitrogen had significant effect only at one location in Kinston in 2014-15. Results suggests that wheat yield is not affected when planted after sorghum (DKS 53-67) compared to corn (DKC 60-67). At Rocky Mount, each year there was no significant effect of pre-plant nitrogen on wheat yield.

WINTER ANNUAL WEED OCCURRENCE IN SMALL GRAINS IN DELAWARE. B. Scott* and M. VanGessel, University of Delaware, Georgetown, DE

ABSTRACT

A three-year IPM project addressing pests in Delaware small grain fields involved surveying a total of sixty-one fields. The weed science program documented weed species diversity and relative density with fall and spring surveys of all fields. The fields were planted to either winter wheat (*Triticum aestivum)* or barley (*Hordeum vulgare*) and included no-till and conventional tillage and fields planted with a drill or seed spread with a spinner and incorporated with a disk or vertical tillage tool. All Delaware counties were represented. Spring surveys were made after all herbicide treatments were applied.

Thirty-eight weed species were identified over the 3-year period. Of those identified, henbit (*Lamium amplexicaule)* and common chickweed (*Stellaria media)* were found on 92 and 88% of the fields, respectively. In the 3-year period, henbit densities ranged from 400 plants/m^2 to a random few plants. No other weed species had such high densities. Annual bluegrass (*Poa annua)* was the next most frequently occurring grass species, documented in 67% of the fields. Field pansy (*Viola bicolor)* and ivy speedwell (*Veronica hederifolia)* were present in greater than 40% of the fields surveyed. Of the fields surveyed, 61% contained 5 to 8 different weed species and less than 1% had a single species present.

Spring assessment timing occurred after herbicide treatments were applied. Weed densities were dramatically lower in the spring and number of species greatly reduced. This sampling resulted in only 2 fields per year having weed densities thought to have impacted yield. Annual bluegrass and confirmed ALS – resistant chickweed infested the two fields in both 2014 and 2015. Two fields in 2016 were infested with annual bluegrass.

Henbit, common chickweed, field pansy, ivy speedwell and annual bluegrass were frequently found in the surveyed fields and it is not surprising they are also difficult to control. Resistance to ALS-inhibiting herbicides has become an issue in small grains in Delaware, specifically with common chickweed. Few effective treatment options are available for control of annual bluegrass in barley.

OVERVIEW OF THE UNIVERSITY OF TENNESSEE WEED DIAGNOSTICS CENTER. J.T. Brosnan*, J.J. Vargas, G.K. Breeden, R.N. Trigiano, and S.L. Boggess, University of Tennessee, Knoxville, TN

ABSTRACT

The Weed Diagnostics Center (WDC) is a new diagnostic arm of the University of Tennessee (UT) Institute of Agriculture that opened in summer 2016. An initiative supported by both the UT Office of AgResearch and UT Extension, the mission of this Center is to provide end-users from across the United States diagnostic tests tailored to weeds of crop production systems as well as turf, ornamentals, and urban landscapes. Serving both the consumer and professional industries, specialists at the WDC incorporate both whole plant and molecular methods to provide a wide range of diagnostic services from basic weed identification to tests confirming herbicide resistance via DNA sequencing. The WDC can currently evaluate weeds for resistance to herbicidal inhibitors of acetyl CoA carboxylase (ACCase), acetolactate synthase (ALS), cellular mitosis, enolpyruvylshikimate phosphate (EPSP) synthase, photosystem II, and protoporphyrinogen oxidase. Work is on going to expand services to test other modes of action and develop more rapid methods of resistance testing. The WDC is also home to an additional suite of services specific to bermudagrasses (*Cynodon* spp.) used on golf course putting greens and fairways. All WDC test results are complemented with research-based control recommendations to promote proper weed management practices in the field. Additional information about the WDC can be found online at www.weeddiagnostics.org, or by contacting weeddiagnostics@utk.edu. More information on the WDC can also be accessed via Twitter (@WeedDiagnostics) and Instagram (@weeddiagnostics).

EFFICACY OF FIVE PREEMERGENCE HERBICIDES AS INFLUENCED BY FIVE DIFFERENT SEEDING INTERVALS BEFORE AND AFTER APPLICATION. C. Marble*, C. Stewart, D. Saha, and A. Chandler, Assistant Professor, Mid-Florida Research and Education Center, University of Florida, Apopka, FL

ABSTRACT

Nursery growers producing container-grown ornamentals must rely heavily on preemergence herbicides, with most making anywhere between 4 and 6 applications per year or more in order to provide season-long control. In many cases, weed escapes can begin to seed before they can be handweeded or sequential preemergence herbicide applications can be made. This creates a situation in which weed seeds are introduced to container substrates at different intervals and small weeds are typically present in various stages of growth before preemergence herbicide applications are made, or seeds are introduced onto substrate surfaces at different intervals soon after application. It is unclear how the efficacy of some of the most commonly used preemergence herbicides would differ depending upon when weed seeds are introduced relative to herbicide application date. The object of this study was to determine how herbicide efficacy would be impacted by seeding date before or after preemergence herbicide applications. These trials were conducted in a greenhouse at the Mid-Florida Research and Education Center in Apopka, FL in 2016. In Sept. 28, nursery containers were filled with a pinebark:peat substrate and standard amendments. Pots were watered in (0.6 cm) and overseeded with 20 seeds of either spotted spurge (*Chamaesyce maculata*) or bittercress (*Cardamine flexuosa*). Additional pots were filled and overseeded every two days until herbicide treatments were applied on Oct. 8. Prior to treatment, 6 additional sets of pots were filled and overseeded as previously described. Herbicide treatments included oxyfluorfen + prodiamine (Biathlon®), pendimethalin + dimethenamid-P (FreeHand®), indaziflam (Marengo®G), oxyfluorfen + pendimethalin (OH2®), and trifluralin + isoxaben (Snapshot®) at 1× their highest labeled rate. After treatments were applied, pots were irrigated using overhead irrigation. One group of pots was overseeded on the day of treatment following irrigation and the five remaining sets of pots were overseeded at 2, 4, 6, 8, and 10 days after treatments were applied allowing for comparison of herbicide efficacy when seeds were sown at 2, 4, 6, 8, and 10 days before (B) or after (A) treatment, as well as seeds that were sown on the day of treatment (0). A non-treated control group was included for each species at each seeding date and shoot fresh weights were used to calculate percent control of each treatment relative to non-treated pots. Biathlon, FreeHand, OH2 and Snapshot provided the lowest percent control ratings when bittercress was seeded either 10 or 8B; no differences were observed in efficacy of these herbicides when seeds were sown at 6B through 10A. Marengo provided less than 57% control of bittercress for seeds sown at 10, 8, 6 and 4B but over 90% control on all other seeding dates. When comparing herbicides on each seeding date, Biathlon generally provided the highest control ratings of any treatment when seeds were sown at 10 and 8B and few differences were observed between herbicides at other seeding dates. For pots seeded with spurge, FreeHand provided the highest percent control ratings of any herbicide when seeds were sown 10 or 8B and few differences were noted between herbicides or in the efficacy of each herbicide between different seeding dates when seeds were sown at 6B through seeds sown at 10A. Overall, few differences were noted between herbicides when seeds were sown between 4B and 10A. Control achieved at 10 and 8B was determined to be due in part to some level of early postemergence control achieved with dimethenamid-p on spurge and oxyfluorfen on bittercress as weeds had already begun to germinate. For pots in which no weeds had germinated at the time of treatment (stages 4B through 10A) few differences were noted in herbicide efficacy for the duration of the trials.

USE OF EH1587 FOR POSTEMERGENCE BROADLEAF WEED CONTROL IN TURF. G.M. Henry*, K.A. Tucker, J.T. Brosnan, G.K. Breeden, and A. Estes, University of Georgia, Athens, GA

ABSTRACT

Field experiments were conducted at the University of Georgia Veterinary College in Athens, GA and the East Tennessee Research and Education Center in Knoxville, TN during the spring and summer of 2016 to evaluate the postemergence control of broadleaf weeds with EH1587. Both cool- and warm-season turfgrasses exhibit tolerance to postemergence applications of EH1587. Research in GA was conducted on bermudagrass (5.1 cm) with a mature infestation of buckhorn plantain (*Plantago lanceolata* L.), while research in TN was conducted on bermudagrass (3.8 cm) with infestations of purple deadnettle (*Lamium purpureum* L.) and hairy bittercress (*Cardamine hirsuta* L.). The experimental design was a randomized complete block with four replications in GA and three replications in TN. Herbicide treatments were initiated in GA on June 3, 2016 and TN on March 4, 2016. Treatments consisted of EH1587 (0.11 kg ai ha^{-1}, 0.21 kg ai ha^{-1}, and 0.31 kg ai ha^{-1}), 2,4-D + mecoprop + dicamba (Triplet) at 1.58 kg ai ha^{-1}, and 2,4-D + triclopyr + dicamba + pyraflufen-ethyl (4-Speed XT) at 1.4 kg ai ha^{-1}. A non-treated check was added for comparison. Herbicides were applied with a CO_2powered backpack sprayer calibrated to deliver 187 L ha^{-1} (GA) or 375 L ha^{-1} (TN). Visual ratings of % broadleaf weed cover were recorded at 0, 14, 28, and 35 days after treatment (DAT). Percent control was determined by comparing % cover on individual rating dates with % cover at trial initiation (0 DAT). EH1587 at 0.21 and 0.31 kg ai ha^{-1} resulted in 57 to 60% purple deadnettle control 14 DAT, while applications at 0.11 kg ai ha^{-1} only provided 47% control. Hairy bittercress control was 57 to 67% 14 DAT, regardless of EH1587 rate. Buckhorn plantain exhibited the highest level of control 14 DAT in response to applications of EH1587. Control was 79 to 89%, regardless of EH1587 rate. Higher control of buckhorn plantain in GA may be attributed to June application timings compared to applications made during March in TN. Triplet at 1.58 kg ai ha^{-1} resulted in 33% purple deadnettle control, 53% hairy bittercress control, and 87% buckhorn plantain control 14 DAT, while 4-Speed XT resulted in 70% purple deadnettle control, 67% hairy bittercress control, and 92% buckhorn plantain control. Excellent broadleaf weed control (98 to 100%) was observed 28 and 35 DAT in response to EH1587, regardless of rate or broadleaf weed species. Control of hairy bittercress and buckhorn plantain with Triplet and 4-Speed XT was also excellent (97 to 100%) 28 and 35 DAT; however, control of purple deadnettle never exceeded 50% in response to Triplet. Purple deadnettle control in response to 4-Speed XT was 90% 28 DAT, but control declined to 77% 35 DAT as weeds began to recover from previous applications.

PREEMERGENCE HERBICIDES IN A PALMER AMARANTH (*AMARANTHUS PALMERI*) MANAGEMENT SYSTEM FOR COTTON. M.C. Askew*, C.W. Cahoon, and A.C. York, Virginia Tech, Blacksburg, VA

ABSTRACT

Glyphosate-resistant Palmer amaranth (AMAPA) is controlled in cotton with well-timed glufosinate applications plus residual herbicides. PRE herbicides are essential components of management systems but they sometimes injure cotton. The objective of this research was to evaluate PRE herbicide combinations and rates as components of an overall management system to control AMAPA while minimizing cotton injury. An experiment was conducted on sandy loam soils in North Carolina at three sites in 2015 and two sites in 2016. Additionally, the experiment was conducted on a loamy sand soil near Suffolk, VA during 2016. Four sites were in conventional tillage and two were planted no-till. No-till sites received glyphosate plus 2,4-D 3 to 4 wk ahead of planting and paraquat PRE. All sites were heavily infested with glyphosate-resistant AMAPA. Twelve PRE herbicide combinations included the following: Warrant + Reflex at 840 + 140, 840 + 210, and 1260 + 280 g ai ha^{-1}; Warrant + Direx at 840 + 560 and 1260 + 560 g ai ha^{-1}; Warrant + Cotoran at 1260 + 1120 g ai ha^{-1}; Reflex + Direx at 140 + 560, 210 + 560, and 280 + 560 g ai ha^{-1}; Reflex + Cotoran at 210 + 1120 g ai ha^{-1}; Brake F16 at 378 g ai ha^{-1}; and Cotoran + Caparol at 840 + 840 g ha^{-1}. A no-PRE treatment was included. Full use rates (g ha^{-1}) of PRE herbicides on the soils in this experiment include the following: Brake F16 (fluridone + fomesafen), 378; Cotoran (fluometuron), 1120; Direx (diuron), 560 if in a combination; Reflex (fomesafen), 280; and Warrant (acetochlor), 1260. Roundup PowerMax (glyphosate) 1260 g ae ha^{-1} + Liberty (glufosinate-amonium) 660 g ai ha^{-1} were applied POST 18 to 35 d after planting when GR-AMAPA averaged 10 cm tall and again 18 to 25 d later. Direx + MSMA at 1120 + 1680 g ai ha^{-1} were directed at layby 14 d after the second POST application. The experimental design was a RCB with three or four replications. The combination of Reflex 280 g ha^{-1} + Warrant 1260 g ha^{-1} , hereafter referred to as the grower standard, controlled AMAPA 99 to 100% at all locations at time of the first POST application. Most PRE combinations and rates controlled AMAPA similar to the standard treatment. Exceptions included Cotoran + Caparol, Reflex 140 g ha^{-1} + Direx 560 g ha^{-1} , and Warrant 840 g ha^{-1} + Direx 560 g ha^{-1} . At 3 of 6 locations prior to POST application, Cotoran + Caparol was 22% less effective than the standard. At 2 of 6 locations, Reflex 140 g ha^{-1} +Direx 560 g ha^{-1} was 17% less effective than the standard. And, at 1 of 6 locations, Warrant 840 g ha^{-1} + Direx 560 g ha^{-1} was 21% less effective than the standard. Following the two POST applications of glyphosate + glufosinate and prior to layby, the standard controlled AMAPA 90 to 100% and averaged 97%. At 4 of 6 locations, AMAPA control averaged 37% less with the no PRE treatment compared with the standard. At 2 of 6 locations, Cotoran + Caparol was 13% less effective than the standard. The layby application masked previously observed differences among the PRE herbicides. All treatments that included a PRE herbicide controlled AMAPA at least 95% after the layby application. However, the value of a PRE herbicide was still evident. Relative to the standard treatment, control in the treatment without PRE herbicides was reduced at all locations by an average of 25%. Injury 14 d after PRE application was usually 10% or less. Injury was most likely to be observed in plots receiving Brake F16 (which contains fomesafen) and Reflex, and injury generally increased as the Reflex rate increased. Seedcotton yield with treatments containing a PRE herbicide did not differ from that of the standard. Yield in treatments without a PRE herbicide was reduced an average of 25%.

JAPANESE STILTGRASS (*MICROSTEGIUM VIMINEUM*) CONTROL IN PASTURES. M.L. Flessner and S.C. Beam*, Virginia Tech, Blacksburg, VA

ABSTRACT

Japanese stiltgrass (*Microstegium vimineum* (Trin.) A. Camus) is a weed that is being increasingly found in pastures across Virginia. Field studies were conducted in 2015 and 2016 in Giles County, Virginia to evaluate POST herbicidal control of Japanese stiltgrass. Treatments were applied on June 8, 2015 and May 13, 2016. Treatments consisted of aminopyralid (Milestone®) at 122 g ae ha^{-1}, metsulfuron (Ally® XP) at 16.8 g ai ha^{-1}, aminopyralid + metsulfuron (Chapparal™) at 130 g ae + 20 g ai ha^{-1}, triclopyr (Remedy® Ultra) at 840 g ae ha^{-1}, glyphosate (Touchdown Total™) at 1165 g ae ha^{-1}, and aminocylcopyrachlor (Method® 240SL) at 175 g ae ha^{-1} in both years. Treatments were arranged in a randomized complete block design with 4 replications and applied with a handheld spray boom calibrated to 140 L ha^{-1}. Visible Japanese stiltgrass control ratings were conducted at 4 and 8 WAT (weeks after treatment) in 2015 and 4, 8, 12, and 16 WAT in 2016 respectively. Ratings ranged from 0 (no control) to 100 (complete necrosis). Data were analyzed in JMP 12 Pro using ANOVA with means separated by Fisher's protected LSD. Data were not pooled across both years due to a significant treatment by year interaction. Metsulfuron and triclopyr resulted in commercially unacceptable control (<60%) in both years. Glyphosate provided 100% Japanese stiltgrass control 4 WAT in both years, but control decreased to 88% 16 WAT. Aminocyclopyrachlor provided good control, 85 and 70% 4 WAT in 2015 and 2016, respectively; control improved to >90% at all other rating dates in both years. Aminopyralid and aminopyralid + metsulfuron provided fair control (>70%) 4 WAT in 2015, however 48% control was observed 4 WAT in 2016. Control improved from aminopyralid containing treatments to >90% at all other rating dates in both years. This research indicates that there are several viable herbicide options available to growers to control Japanese stiltgrass POST in pastures with glyphosate, aminocyclopyrachlor, aminopyralid, and aminopyralid + metsulfuron. Future research may assess the tolerance of pasture forage grasses to these herbicides as well as evaluate PRE herbicide options.

EFFECTS OF SUB-LETHAL DICAMBA RATES APPLIED TO SOYBEAN AT THE REPRODUCTIVE GROWTH STAGE. A.M. Growe* and W.J. Everman, NCSU Crop Science, Raleigh, NC

ABSTRACT

Dicamba-resistant crop varieties have the potential to become utilized in North Carolina as a tool to control glyphosate-resistant weeds. Current soybean cultivars, commonly glyphosate or glufosinate-resistant varieties, are highly susceptible to dicamba, and there is growing concern of off-site movement of this broadleaf herbicide to sensitive cultivars. Tank contamination, wind drift, and volatility of dicamba can cause injury and reduce soybean yields. To date, there has been little information reported on soybean varietal responses to sub-lethal doses of dicamba.

The objective of this study was to evaluate the effects of sub-lethal rates of dicamba on five maturity group V and five maturity group VI soybean cultivars at the reproductive growth stage. Effects of dicamba were determined by collecting visual injury ratings, height reductions and yield. Experiments were conducted in Lewiston-Woodville, Kinston, and Rocky Mount, North Carolina during 2015 and 2016. Five soybean varieties were treated with dicamba at 1.1, 2.2, 4.4, 8.8, 17.5, 35, and 70 g ae ha (1/512 to 1/8 of the labeled use rate for weed control in corn) during the R2 growth stage. Experiments were conducted using a factorial arrangement of treatments in a randomized complete block design, with two factors being dicamba rate and soybean cultivar. All data were subjected to analysis of variance and means were separated using Fisher's Protected LSD at p= 0.05.

Analysis showed a wide range of visual injury and height reduction 2 and 4 WAT for all 10 varieties. Higher levels of height reductions and injury were associated with increasing dicamba rates. When data was pooled across all soybean varieties in each maturity group, height reductions 4 WAT ranged from 2-35% and 4-30% for maturity group V and VI respectively. Injury 2 WAT ranged from 20-65% for group V and 18-67% for group VI soybeans. Statistical analysis revealed a location and variety interaction for injury ratings. This data suggests the potential for a varietal response to sub-lethal rates of dicamba. Dicamba effects on soybean yield have not yet been analyzed.

THE IR-4 PROJECT: UPDATE ON WEED CONTROL PROJECTS (FOOD USES). M. Arsenovic*, J. Baron, D. Kunkel, and R. Batts, IR-4 Project, Princeton, NJ

ABSTRACT

The IR-4 Specialty Crop Program is a publicly funded program that develops and submits regulatory data for the registration of pest control products on specialty crops. IR-4 has a long history of providing herbicide registrations for specialty or minor crop growers. In 2016, IR-4 data was used to support a number of new herbicide registration. These included new registration for carfentrazone-ethyl, clethodim and penoxulam in many crops, and pronamide in leaf lettuce. EPA issued Notices of Filling for use of isoxaben ethofumesate, indaziflam, and pyroxasulfone. Herbicide submissions were made to the EPA in 2016 for the use of clopyralid, 6-benzyladenine, clethodim, prometryn, rimsulfuron, and quinclorac in many crops. In addition, IR-4 submitted final report to the Registrant for label change for the use of hexazinone in blueberry. The 2017 IR-4 research plan will include about 40% herbicide projects. The IR-4 Project will remain a responsive and efficient organization that supports US stakeholders by facilitating U.S. regulatory approvals for crop protection tools. IR-4 will continue with a research focus on low risk pesticide registrations that support integrated pest management systems. There will be more emphasis on efficacy and crop safety data to determine products that can manage hard to control pests and mitigate invasive species. There will be continued support for products that can be used for organic growers, as well as cutting edge biotechnology products. Other areas expected to have greater importance will be in the area of international harmonization to assist in exports of U.S. grown specialty crops and removing pesticides as a technical trade barrier.

PESTICIDE APPLICATION MANAGER, A PROGRAM FOR APPLICATORS. B.F. McDonnell*, National Park Service, Bushkill, PA

ABSTRACT

The Pesticide Application Manager (PestAppMan) is a free Windows®-based program designed to assist applicators with many common calculations and records keeping tasks. Its components include two pesticide mixing modules, a pesticide daily use log, a boom calibration worksheet, a simple calibration calculator, a labels and documents viewer, and a calculator to convert between pesticides with different concentrations of a common active ingredient.

GERMINATION TIMING OF TWO HAIRY VETCH VARIETIES IN VIRGINIA AND IMPLICATIONS FOR WEEDINESS. K.B. Pittman*, M.L. Flessner, C.W. Cahoon, J.N. Barney, and T. Hines, Virginia Tech, Blacksburg, VA

ABSTRACT

There has been a renewed interest in the use of cover crops that stems from a growing awareness of sustainable agricultural practices and an increase in cases of herbicide resistant weeds. Hairy vetch (*Vicia villosa* Roth.), a leguminous cover crop species, can become a weed in subsequent cash crop production. Research was conducted to evaluate the germination patterns of two hairy vetch varieties at different planting depths over a one-year period.

Studies were conducted in three locations: Blacksburg, Blackstone, and Painter, Virginia, respectively. These experiments were set up in a randomized complete block design with 6 replications of a 2^3 factorial. Factors were: two hairy vetch varieties, burial depth, and whether the boxes were covered or uncovered. The two varieties were Groff and Purple Bounty, a late flowering and early flowering variety, respectively. Seeds were either placed on the soil surface, to simulate seed being flown on, or buried to a depth of 2.5 cm, to simulate a drill depth. The last factor being whether the box was covered or uncovered. This was to prevent vertebrate predation on the seeds. The experiments were started on October 20, 2015, November 3, 2015, and November 6, 2015 in Blackstone, Blacksburg, and Painter, respectively. Wire mesh boxes (0.3m by 0.3m by 0.18m) were constructed to place in the field and 500 seeds were placed in each box. Boxes were monitored for germination on a two week basis during the fall and spring and monthly during the winter and summer months. Once a seed was considered germinated, it was removed from the box. Plants were considered to have germinated once a radicle emerged. Data were analyzed using JMP 11 and were subjected to ANOVA and effects were considered significant when $p < 0.05$ followed by a means comparison across treatments using Fisher's protected LSD ($p<0.05$).

Groff had a higher germination rate than Purple Bounty across all locations one, six, and twelve months after planting (MAP). Seeds sown at the soil surface had a higher germination rate rather than those buried in Blacksburg and Painter but the opposite was observed in Blackstone. Whether or not the boxes were covered was not significant in germination rate at the Blackstone or Painter locations. In Blacksburg, covered boxes had a higher germination rate than those that were uncovered. At 12 MAP in Blackstone, 99% of Groff seeds had germinated and 67% of Purple Bounty seeds had germinated. In Blacksburg, 85% of Groff seeds had germinated, while only 57% of Purple Bounty seed had germinated at 12 MAP. Inversely, 15% and 43% of Groff and Purple Bounty seeds, respectively, have not germinated to date in Blacksburg. These seeds may still be viable in the seedbank and germinate. Overall, these data indicate that hairy vetch seed may remain dormant after initial seeding which can result in the potential for these plants to become weeds in subsequent crops. Future research will consist of continuing germination observations and determination of the viability of any seed remaining.

BICYCLOPYRONE: MAJOR LEAGUE WEED CONTROL IN MINOR LEAGUE CROPS. L. Smith*, C. Dunne, G. Vail, D. Bruns, T. Beckett, and S. Howard, Syngenta, King Ferry, NY

ABSTRACT

Bicyclopyrone is a newly registered HPPD-inhibiting active ingredient for control of dicot and some grass weeds. Bicyclopyrone is one of the four active ingredients in Acuron herbicide which was registered for sales in corn in 2015. Syngenta is evaluating the potential for expanding bicyclopyrone use into minor/specialty crops where options for weed control are limited. Since 2013, University and Syngenta trials evaluated both PRE and POST bicyclopyrone applications for crop tolerance and weed control in minor crops including carrot, horseradish, hops, onion, rosemary, strawberry, sweet potato, timothy, watermelon, and ornamentals.

USING RAINOUT SHELTERS TO TEST DROUGHT EFFECTS ON IVYLEAF MORNINGGLORY COMPETITION IN CORN. A. DiTommaso*, K.M. Averill, S.H. Morris, R.J. Richtmyer III, and M.C. Hunter, Cornell University, Ithaca, NY

ABSTRACT

Climate change can exacerbate weed population growth and spread. Consequently, testing for climate change impacts on crop production is becoming an essential research avenue in weed science. While controlled environment experimentation allows for a wide range of finely tuned temperature, moisture, and CO_2 manipulations, field experimentation offers a more representative setting to investigate climate change effects on weed growth in actual cropping systems. Here we focus on creating short-term drought conditions with rainout (i.e., rain exclusion) shelters that are established after planting and early season management practices. Shelters can be affordably constructed with steel electrical conduit and durable, high clarity plastic. Using a total of 20 rainout shelters 2.7 m wide by 3 m long by 3 m tall, complete with gutters and drainage tubing to divert water away from plots, the effect of short-term drought on ivyleaf morningglory (*Ipomoea hederacea* Jacq.) (IMG) competition was tested in glyphosate-resistant corn (*Zea mays* L.) during the 2016 growing season in Ithaca, NY. In early June, IMG seedlings were transplanted within rows of corn at five densities (0, 1, 2, 4, 8 plants m^{-2}). IMG seedlings were approximately the same size as corn seedlings. Corn was planted in late May at a density of 71,660 plants ha^{-1} with 76 cm row spacing. Other weeds were controlled through a combination of glyphosate applied once prior to transplanting IMG, inter-row cultivation in late June, and weekly hand-weeding throughout the experiment. Due to an extremely dry June, non-drought treatment plots were irrigated with approximate 2.5 cm watering events weekly prior to establishing rainout shelters in late July and until crop and weed harvest in mid-September. Soil volumetric water content was measured twice weekly during the growing season using fixed-in-place, time-domain reflectometer probes. Soil water content was lower in drought plots with rainout shelters (0.086 ± 0.003 cm^3 water cm^{-3} soil) compared with non-drought plots without rainout shelters (0.116 ± 0.003 cm^3 water cm^{-3} soil) ($p < 0.001$). A negative correlation was found between IMG density and dry corn silage (slope = -0.08 ± 0.04; $p < 0.05$), but not fresh corn silage. Preliminary results suggest that silage yield declined as IMG density increased, however the drought treatment did not translate into significant differences in corn silage yields or in IMG biomass between drought and non-drought plots. An important methodological takeaway from the 2016 experiment is to reinforce the shelter legs and footings, as the bases of the legs were partially buried due to wind action, which caused shelter instability and required repeated maintenance. In 2017, the experiment will be repeated in the same field with reinforced shelter legs and additional irrigation of non-drought plots if necessary. In the event of similar ambient drought conditions as in 2016, increased irrigation quantities in non-drought plots in 2017 would be expected to produce a stronger treatment response.

COMPARISONS OF SINGLE VERSUS SPLIT POSTEMERGENCE MESOTRIONE APPLICATIONS FOR SPREADING DOGBANE CONTROL IN WILD BLUEBERRY FIELDS. D.E. Yarborough and J. D'Appollonio*, University of Maine, Orono, ME

ABSTRACT

Spreading dogbane (*Apocynum androsaemifolium*) is a major weed pest in wild blueberry (*Vaccinium angustifolium*) fields, and is difficult to control with many of the industry's currently registered herbicides. In spring 2015 we initiated a trial at the University of Maine's Blueberry Hill Farm to examine the effect of mesotrione (Callisto) and rimsulfuron (Matrix) on dogbane control. Dogbane was sprayed post-emergence at the 3 to 5 leaf stage to four 1 by 2 m plots either once at Callisto 6 oz/a or Matrix 4 oz/a or in a split application at 3 oz/a or 2 oz/a applied twice, both alone or in combination for a total of six herbicide treatments and also included an untreated check and the Farm's herbicide treatment of hexazinone 2 lb/a+terbacil 2 lb/a+diuron 1.6 qt/a pre-emergence and Callisto 6 oz/a+clethodim 6 oz/a post-emergence. In 2015, the Callisto and Callisto+Matrix tank mixes resulted in higher dogbane injury and lower cover overall, which suggested that Callisto was more effective in controlling dogbane. The split Callisto treatment was the most effective in controlling dogbane and resulted in the lowest dogbane cover and greatest dogbane injury along with the split Callisto+Matrix treatment, but considering the two were equal had equal blueberry cover and phytotoxicity, the addition of Matrix did not improve the control and was considered unnecessary. Wild blueberry cover and dogbane control was assessed in the crop year June 2016. There were no significant differences among treatments for wild blueberry or dogbane cover, but the most effective dogbane reduction was observed in the single and split Callisto treatments. Two Callisto applications resulted in better residual dogbane reduction compared to one application, regardless of whether it was combined with Matrix indicating the addition of Matrix was ineffective for dogbane control. In a trial initiated in spring 2016 in a commercial pruned field dogbane was sprayed post-emergence with Callisto at either 2 oz/a or 3 oz/a to six 1 by 4 m plots split in half with one half receiving pre-emergence hexazinone (Velpar). Once the blueberry plants emerged, dogbane emergence and growth were tracked weekly and the plots were sprayed when dogbane reached the 3 to 5 leaf stage at approximately two week intervals for a total of three Callisto applications and four evaluations. All treatments were compared, as well as Velpar vs no Velpar for the main Callisto treatments. Although the Velpar combinations and Callisto 3 oz/a alone almost eliminated dogbane and no new seedlings were observed at the last evaluation in July, dogbane was not completely controlled by any treatment. Some dogbane stems which appeared dead at the third evaluation showed regrowth of lateral leaves in July. The results of Velpar vs no Velpar indicate that the addition of Velpar slightly increased dogbane control and injury. Dogbane control and wild blueberry yield will be assessed in 2017.

IMPROVING GRANULAR HERBICIDE APPLICATION UNIFORMITY IN CONTAINER NURSERIES. J. Neal* and A. Barker, North Carolina State University, Raleigh, NC

ABSTRACT

Granular herbicide applications in container nurseries have been shown to be highly variable, with up to 250% variation in dose applied to containers within a block of plants. The standard application device is a hand-held, hand-cranked (belly-grinder) spreader. Tests were conducted to identify spreader settings and operator variables that affect overall distribution patterns and how these factors could be altered to improve granular herbicide application uniformity. Two spreaders were chosen because of their prevalence in the nursery industry: the Warren T-7 II and the Solo 421S. Experiments were conducted using inert granules for the herbicides FreeHand (irregular clay granules) and Biathlon on Verge (roughly spherical manufactured granules); and three mesh sizes of Biodac (irregular granules of recycled paper). Using plastic collection bins placed 2-feet apart, granule distribution patterns were measured for each spreader and granule type for six replicate passes. Granules in each collection bin were weighed, and data were analyzed using the SPREADEREZ program. Additionally, spreader patterns using different directional rudder positions and volume output (rate) settings were evaluated in the same manner.

The center directional rudder position produced the most uniform pattern. Left and right rudder positions produced unacceptably high variability in granule distribution. Changes in the spreader rate setting also affected pattern uniformity. The most uniform pattern was observed at the highest setting. At lower settings the spreader distribution was skewed to the right, resulting in high coefficients of variability (CV). Using data collected in these tests, distribution patterns at different swath widths were simulated using SPREADEREZ. For all granules tested, the lowest CV was achieved with swath widths of 8 to 10 feet. The CV increased dramatically at wider swath widths. At the optimum swath width of 8 to 10 feet the CV for granule distribution was acceptable only when using the highest output setting. The walking pattern of the operator can be altered to compensate for skewed distributions at lower dose settings. Applications in parallel passes walking in the same reduced the CV; however, this is generally not practical. A preferable method of application is to calibrate for half-dose applications then walk "up and back" on the same path. This resulted in the most uniform distribution.

The results of this research and prior reports were integrated to provide guidelines for granular spreader usage for herbicide applications in container nurseries. This fact sheet, AG-826, is available from NC State University Extension Resource Catalog, https://content.ces.ncsu.edu/.

RATTLESNAKES-YES, TREES-NO: QUAKING ASPEN SUPPRESSION IN THE JENNINGS PRAIRIE WITH PRESCRIBED FIRE, MOWING, AND FALL HERBICIDE APPLICATIONS. A. Gover* and W. Taylor, Penn State University, University Park, PA

ABSTRACT

The Jennings Prairie is one the easternmost remnants of prairie featuring dense blazingstar (*Liatrus spicata* (L.) Willd.), and is also hosts the Federally Threatened Eastern Massasauga rattlesnake (*Sistrurus catenatus catenatus,* [Rafinesque, 1818], hereafter EMR). Maintenance had evolved to a regimen featuring prescribed fire every two years, supplemented with mowing in late spring or winter. Vegetation management operations are conducted in a manner to prevent injury to the EMR or its hibernacula. Woody vegetation, particularly suckering species such as quaking aspen (*Populus tremuloides* L.), gray dogwood (*Cornus racemosa* L.), and American hazel (*Corylus americana* Walter) were being kept at a small size, but density was increasing. The existing management quadrants were subdivided to incorporate regimens comparing late spring mowing or late season herbicide treatment in 2- or 4-year fire rotations in the fall of 2014, with the intent to complete each rotation, and then determine if one approach should be adopted for the entire prairie. For each treatment regimen, there was one operation per year, so that mowing or herbicide application occurred in each year there was no prescribed fire. Within each plot, three permanent 9 m^2 sub-plots were established, with woody stem counts taken annually in the center 1 m square, and herbaceous cover assessed as percent cover and dominant species in the entire sub-plot. Late spring mowing was conducted using a walk-behind sickle bar mower cutting at approximately 20 cm to avoid injuring EMR. Herbicide treatments were conducted in September, prior to onset of fall coloration, using the "glove of death" stem wipe method, applying triclopyr (ester formulation) at 90 g/L in oil to a 15 to 20 cm band on all woody stems using a cloth glove wetted with the herbicide mixture. This method was selected because it was easier than applying glyphosate to foliage in a similar manner, and the injury reduction to non-target forbs mitigated the labor-intensiveness. At the initiation of the treatments, the overall average stem density for quaking aspen, total suckering species, and total woody species was 9.0, 12.2, and 18.0 stems/m^2. The treatment sequence was complicated in 2016 when prescribed burns were not conducted due to lack of agency consensus to proceed, affecting a replicate of both the 2- and 4-yr rotations. Therefore, the woody species stem density analysis only considered mowing vs. herbicide. Data for 2016 was converted to percent of initial (2014) stem counts, and reported as quaking aspen, all suckering species, all woody species, and herbaceous species cover. Quaking aspen stem density was 67 and 23 percent of initial for mowing and herbicide, respectively, but was not significantly different. Results for total suckering species and total woody species counts were significantly different, with mowing vs. herbicide comparisons of 110 vs. 23, and 101 vs. 27 percent, respectively. Herbaceous cover averaged 59 percent for mowing, and 79 percent for herbicide treatment, but was not significantly different.

WEED SUPPRESSION BY COVER CROPS MIXTURES USING INTRA- AND INTERSPECIFIC DIVERSITY. E. Reiss*, L.E. Drinkwater, and M. Ryan, Cornell University, Ithaca, NY

ABSTRACT

Cover crops are commonly used to suppress weeds prior to establishment of a main crop, along with other benefits. Based on natural ecosystem research, which has demonstrated a positive relationship between diversity and productivity, we expected that more diverse cover crop mixtures would produce more biomass and consequently reduce weed biomass. Diversity can be measured at a number of different levels. Here we looked specifically at three levels: functional diversity (legumes: nitrogen-fixing, grasses: non-N fixing), interspecific diversity (multiple species), and intraspecific diversity (multiple cultivars within a species). During the overwinter cropping season, in two consecutive years at Musgrave Research Farm in Aurora, NY, we established an experiment with a randomized complete block, replacement series design. We created a diversity gradient with 49 treatments using 6 annual cover crop species (3 legumes, 3 grasses) and multiple cultivars of each species. Biomass and plant counts for all cover crop species and weeds were collected prior to termination of the cover crops. Results show that mixtures of multiple cultivars suppress weeds better (as biomass) than single cultivars alone. Additionally, species polycultures divert more productivity from weeds compared to monocultures. Cover crop biomass increases with increasing number of species in the mixture. Functional group effect depends on identity, but legumes always benefit from mixture with grasses in terms of cover crop biomass and weed suppression. These results suggest that mixtures with different levels of diversity can be beneficial in promoting biomass production and weed suppression.

CONFIRMATION OF GLYPHOSATE, ALS- AND PPO-RESISTANT COMMON RAGWEED IN NORTH CAROLINA. B. Schrage* and W.J. Everman, North Carolina State University, Raleigh, NC

ABSTRACT

Beginning in 2014, soybean growers in northeastern North Carolina have had increasing difficulty controlling common ragweed with available PPO-inhibiting herbicides. These herbicides included: aryl triazinones, diphenylethers, and N-phenylphthalimides-many of which are labeled for *Ambrosia* control in soybeans. The presence of previously confirmed ALS- and glyphosate-resistant biotypes in the state prompted the need to conduct a bioassay to determine if common ragweed had developed multiple resistance.

In 2015, a putative-resistant population from Currituck County along with a control population from Edgecombe County were screened against several PPO- and ALS-inhibiting herbicides. Application rates for the susceptible population were 1/32, 1/16, 1/8, ¼ , ½ and 1X the labeled field rate. Putative-resistant individuals were treated with 1, 2, 4, 8, 16, and 32x the labeled field rate. It was determined that Edgecombe County biotype had indeed inherited resistance to PPO- and ALS-inhibiting herbicides, allowing individuals to survive a 32x application of lactofen. A second screening involving additional herbicides and a larder range of rates for a multitude of populations is currently being conducted. PPO-resistant common ragweed in North Carolina confirms the first agronomic, dichotomous weed species to exhibit resistance to three modes of action within the state.

SEED SHATTERING OF SIX PREVALENT WEED SPECIES IN NORTH CAROLINA. T.A. Reinhardt*, D. Copeland, and W.J. Everman, North Carolina State University, Raleigh, NC

ABSTRACT

As herbicide resistant weeds become more common, harvest weed seed control has gained interest as part of a weed resistance management plan. However, the success of these methods for control require weeds retain seed until harvest in order for seeds to be removed with the crop. In this study, we examine the timing of seed dispersal of six prevalent weed species in North Carolina: Palmer amaranth, common ragweed, sicklepod, common cocklebur, broadleaf signalgrass, and Texas panicum. Four trays 51.4cm x 40cm x 6.3cm were anchored around each plant in October. Fallen seeds were vacuumed from the trays and collected for counting each week until the end of typical soybean harvest in North Carolina. At the last collection, plants were cut at the soil surface, threshed, and seeds counted. This data will help determine which species have the best chance to be controlled by harvest weed seed control.

REMOTE SENSING APPLICATIONS FOR PALMER AMARANTH DETECTION. J.T. Sanders* and W.J. Everman, North Carolina State University, Raleigh, NC

ABSTRACT

The proliferation of unmanned aerial vehicles (UAVs) into the mainstream consumer market has generated questions about their utility in commercial agricultural production. Aerial imagery has been previously used to diagnose crop status with respect to such factors as nutrient and water availability as well as pathogen mapping, but the acquisition of this imagery has been subject to high cost, long revisit times and concerns about image resolution. Given their small size, ease of use and extraordinary maneuverability, UAVs may indeed provide a cost effective and efficient means to obtain aerial imagery from crop fields to aid in making management decisions. In 2015, a field study was established to examine the ability of a commercially available UAV equipped with an integrated RGB camera to distinguish *Amaranthus palmeri* (palmer amaranth) from a soybean crop at altitudes of 50, 100 and 150 feet throughout the growing season. Imagery was then subject to a supervised classification utilizing a maximum likelihood separation. Increasing altitude generally increased overall classification accuracy, while progression through the season resulted in a degradation of this metric

HORSEWEED (*CONYZA CANADENSIS*) CONTROL WITH BURNDOWN HERBICIDES AFTER WINTER WHEAT HARVEST. S.C. Haring*, M.L. Flessner, and K.B. Pittman, Virginia Tech, Blacksburg, VA

ABSTRACT

Many summer annual weeds, including horseweed (*Conyza canadensis* (L.) Cronq.), that affect soybean (*Glycine max* (L.) Merr.) production are present at the time of winter wheat (*Triticum aestivum* L.) harvest when planted in a double crop rotation. Harvest equipment, especially the combine's cutter bar, has the potential to reduce leaf area, which may impact efficacy of burndown herbicides to control horseweed ahead of soybean planting. Field and greenhouse studies were conducted in Blacksburg, Virginia in 2016 to investigate how horseweed that has been cut by a combine respond to herbicides, including glyphosate alone and in combination. The field experiment was set up in a randomized complete block design with three replications. Herbicide and adjuvants used included glyphosate (RoundUp PowerMAX®) at 1.26 kg ae ha^{-1}, dicamba at 0.56 kg ai ha^{-1}, 2,4-D amine at 1.12 kg ai ha^{-1}, paraquat at 0.84 kg ai ha^{-1}, glufosinate (Liberty®) at 0.59 kg ai ha^{-1}, metribuzin (TriCor®) at 0.315 kg ai ha^{-1}, nonionic surfactant (NIS) at 1% v v^{-1}, and ammonium sulfate (AMS) at 3.36 kg ha^{-1} in the following treatment combinations: glyphosate alone, 2,4-D amine alone, dicamba with NIS, glyphosate with 2,4-D amine, glyphosate with dicamba and NIS, paraquat with NIS, 2,4-D amine with paraquat and NIS, dicamba with paraquat and NIS, glufosinate with AMS, metribuzin alone, metribuzin with 2,4-D amine, metribuzin with dicamba, and metribuzin with paraquat, as well as a nontreated check. Herbicides were applied with a backpack sprayer at 140 L ha^{-1} to glyphosate-susceptible horseweed that had been cut with hedge trimmers at approximately 0.46 m from the ground, to simulate the height at which a combine would cut at wheat harvest. Visual control ratings were taken 2, 4, and 6 weeks after treatment (WAT). The greenhouse study was factorial in nature, with factor A being horseweed size and factor B being herbicide treatment. Levels for factor A were horseweed in rosette stage, bolting, and cut to 0.46 m. Factor B used the same herbicide treatments as the field study, less the treatments containing metribuzin. Visual control ratings were taken 1, 2, 4, and 6 WAT, and plants were harvested, dried, and weighed for biomass immediately following the final rating. Data were analyzed using ANOVA with a multiple comparison test using Fisher's Protected LSD at a significance level of $p<0.05$. In the field experiment, all treatments except those containing paraquat produced commercially unacceptable control 2 WAT. By 6 WAT, all treatments produced ≥85% control except for glyphosate alone, glufosinate with AMS, and metribuzin alone, which resulted in 60%, 53.3%, and 15%, respectively. In the greenhouse study, glyphosate alone, 2,4-D amine alone, dicamba with NIS, glyphosate with 2,4-D, and glyphosate with dicamba and NIS produced commercially unacceptable control results and higher final biomasses than other herbicide treatments, across all three horseweed lifecycle stages. Only treatments containing paraquat or glufosinate consistently produced >80% control from 2 WAT on, and they did so across all horseweed lifecycle stages. Additionally, these treatments had lower final weed biomasses than other herbicide treatments across all horseweed lifecycle stages. These results suggest that commercially acceptable horseweed control can be attained with herbicide applications after plants have been cut in a winter wheat harvest. Additionally, the most efficacious herbicide mixtures, those including paraquat, were effective at controlling horseweed across various growth stages, as may be present in field settings at the time of wheat harvest. Further research is needed to assess how these herbicide combinations might affect other important herbicide-resistant weed species at winter wheat harvest.

QUANTIFICATION OF SALVAGE APPLICATIONS WHEN MANAGING MULTIPLE-RESISTANT PALMER AMARANTH IN LIBERTY-LINK SOYBEANS. D. Copeland* and W.J. Everman, North Carolina State University, Raleigh, NC

ABSTRACT

Palmer amaranth (*Amaranthus palmeri*) is the most problematic weed in North Carolina soybean production. Due to glyphosate-resistant (GR) Palmer amaranth, increased selection pressure on PPO-inhibiting herbicides have selected for PPO-resistant Palmer amaranth biotypes throughout the Mid-South and the Carolinas. Currently growers rely heavily on glufosinate to control this weed. Timely applications are critical for adequate weed control with glufosinate, but are not always feasible due to unforeseen circumstances. Due to restrictions on new technologies, tank-mixes that include dicamba POST are currently considered off-label applications. Therefore, glufosinate included within a tank-mix is the only effective, labeled option for POST control of multiple-resistant Palmer amaranth. The objective of this study was to evaluate how tank-mixes of PPO-herbicides in combination with glufosinate perform in salvage applications on multiple-resistant Palmer amaranth.

The experiment was conducted in 2016 in Cheraw, South Carolina in a grower field with natural infestations of PPO-resistant-GR Palmer amaranth. Herbicide applications were initiated when Palmer amaranth plants reached 15 to 20 cm. and 25 to 30 cm. in height. Treatments utilized in this experiment included: glufosinate-ammonium (740 g ai ha^{-1}) alone, fomesafen (395 g ai ha^{-1}) alone, lactofen (280 g ai ha^{-1}) alone, acifluorfen (420 g ai ha^{-1}) alone, flumiclorac (60 g ai ha^{-1}) alone, glufosinate-ammonium + fomesafen (740 + 395 g ai ha^{-1}), glufosinate-ammonium + lactofen (740 + 280 g ai ha^{-1}), glufosinate-ammonium + acifluorfen (740 + 420 g ai ha^{-1}), glufosinate-ammonium + flumiclorac (740 + 60 g ai ha^{-1}), fomesafen + lactofen (395 + 280 g ai ha^{-1}), fomesafen + acifluorfen (395 + 420 g ai ha^{-1}), and fomesafen + flumiclorac (395 + 60 g ai ha^{-1}). Each treatment contained methylated seed oil (MSO) at 0.5 % v/v and 1 % v/v. A non-treated check was included. The experimental design was a RCBD with four replications.

Regardless of timing, rate of MSO had no effect on weed count reductions or weed biomass for any herbicide treatment combination. At the 15 to 20 cm timing, Palmer amaranth densities and biomass were greater 3 weeks after treatment (WAT) when treated with lactofen or flumiclorac alone compared to all other treatments. However at the 25 to 30 cm, any tank-mix excluding glufosinate or glufosinate applied alone resulted in significantly greater weed densities and biomass 3 WAT. This research demonstrates that salvage control of multiple-resistant Palmer amaranth is possible at the 15 to 20 cm timing where a tank-mix of glufosinate + fomesafen will reduce weed densities (93 %) and biomass (99%). However, applications applied later will only reduce Palmer amaranth densities and biomass by 76 and 88%, respectively.

EMERGENCE OF 6 GRASS SPECIES IN NORTH CAROLINA. W.J. Everman and M.T. Schroeder*, North Carolina State University, Raleigh, NC

ABSTRACT

Knowing which weed species are problematic and when they emerge are necessary in order to set up an effective weed management plan. This data is especially important when applying herbicides with no residual. In this study barnyard grass, broadleaf signal grass, fall panicum, goosegrass, large crabgrass, and texas panicum were all sown into 36 inch sorghum rows. Emergence timing as well as end of season heights, tiller count, and seed count were taken to document biomass differences in large verses small populations. Peak emergence was at similar timings for all species except fall panicum, which had peak germination one week later. The preliminary data showed that barnyardgrass, broadleaf signal grass, and large crabgrass have the ability to germinate many seedlings over a 3 week period.

BENEFITS AND CHALLENGES OF PLANTING CORN AND SOYBEAN INTO LIVING COVER CROPS IN PENNSYLVANIA . H.K. Myer*, H.D. Karsten, J. Tooker, W. Curran, and S. Duiker, Penn State University, State College, PA

ABSTRACT

Planting into living cover crops, or "planting green," is a novel practice used by some Pennsylvania farmers to conserve soil and improve soil health, and manage soil moisture. The typical practice in no-till systems involves killing cover crops with herbicide about two weeks prior to cash crop planting. By delaying termination until after cash crop planting, the benefits of cover crops can be extended, but the increased cover crop growth can also pose management challenges. We hypothesized that planting green would increase soil residue cover, reduce soil moisture in spring and conserve soil moisture later in the growing season, increase beneficial arthropod populations and predation, and reduce slug herbivory on cash crops. Field experiments were conducted at two Penn State research stations and three cooperating farm locations in central and southern Pennsylvania in 2015 and 2016 to quantify the effects of termination timing and cover crop species on cover crop biomass, cash crop populations, soil moisture and temperature, slug and beneficial arthropod populations, and cash crop yield. In both years cover crop biomass increased approximately two-fold biomass when cover crops were terminated at planting compared to the typical "early-termination," practice, resulting in reduced cash crop populations in some site-years. Planting into a living cover crop reduced soil moisture at planting in spring and conserved soil moisture later in the summer, while soil was cooler throughout the growing season in late-terminated treatments. Slug populations were variable and in some cover crop species, delaying cover crop termination increased beneficial arthropod abundance in corn. Overall, soybeans were able to compensate for reduced plant populations and environmental stresses, and yield was not influenced by cover crop termination timing in most site-years. However, achieving similar corn yields when planting into living cover crops compared to early-terminated cover crops was complicated by residue management, especially in 2015, a year with dry spring conditions. When spring weather is dry, planting green is generally not recommended, although planting soybeans into living cover crops in 2015 did not reduce yields. Planting soybeans into live cover crops appears to be a promising practice for farmers who want to extend the benefits of their cover crops.

IMPACT OF HERBICIDE FORMULATION, MULCH DEPTH AND ACTIVATION MOISTURE ON WEED CONTROL EFFICACY IN LANDSCAPE PLANTING BEDS. D. Saha*, C. Marble, B.J. Pearson, G.E. MacDonald, D. Odero, and H.E. Perez, Graduate Research Assistant, Mid-Florida Research and Education Center, University of Florida, Apopka, FL

ABSTRACT

There are currently no label recommendations for preemergence herbicide use in mulched landscape beds. One of the most important aspects of preemergence herbicides is that they must be incorporated into the soil with rainfall or irrigation soon after application to be "activated". It is unknown if more or less moisture is needed if an area is mulched, nor is there information available on how different mulch materials and herbicides interact. Objectives of this study were to assess impact of herbicide formulation, mulch type and depth, and activation moisture on weed control in planting beds. This research was conducted at the Mid-Florida Research and Education Center in Apopka, FL in 2016. Nursery containers (946.35 ml) were filled with standard substrate and amendments. After filling, pots were overseeded separately with equal amounts of either crabgrass (*Digitaria sanguinalis*), garden spurge (*Chamaesyce hirta*) or eclipta (*Elipta prostrata*). Following seeding, three different types of mulch including pinestraw, pinebark or hardwood were added on top of each container at depths of 0, 2.54 or 5.08cm. Granular or liquid formulations of indaziflam, prodiamine, or dimethanamid-P + pendimethalin were applied on August 17 at label rates. Liquid formulations were applied with a CO_2 backpack sprayer calibrated to deliver 178 liters per hectare while granular formulations were applied using a hand-shaker. Each pot received either 1.27 cm, 2.54cm, or 5.08cm of irrigation after treatments were applied in order to assess the impact of activation moisture, mulch depth, and herbicide formulation on weed control. Pots were kept inside a greenhouse and irrigated 0.5 cm per day throughout the trial. The experiment consisted of a factorial treatment arrangement of the two herbicide formulations, three types of mulch, three depths, and three after treatment, one-time irrigation levels. Data collection included weed counts at 30 and 60 days after treatment (DAT). At 60 DAT, shoot fresh weights were determined individually for each species. Data were subjected to ANOVA with the use of the PROC GLM procedure in SAS® and Fisher's Least Significance Difference Test was used to compare between individual means of experimental variables. All differences considered significant at $p = \leq 0.05$ and each weed species was analyzed separately. None of the experimental response variables were impacted by after-application, one-time irrigation level. Mulch type was not a significant main effect in either crabgrass or garden spurge. Pinestraw and hardwood chips provided greater control of eclipta when compared to pinebark. For all three weed species, mulch depth of 5.08 cm provided greater control than depths of 0 or 2.54cm. Spray-applied formulations of indaziflam and prodiamine provided greater control than their respective granular formulations for controlling eclipta and crabgrass, respectively. Greater spurge control was achieved with granular formulations of dimethenamid-p + pendimethalin. Based upon results from this trial, mulch depth and herbicide formulation will impact weed control efficacy to a greater degree than activation moisture or mulch type.

EFFECTS OF COVER CROP MANAGEMENT AND TILLAGE ON WEED SUPPRESSION, SOIL HEALTH, AND PROFITABILITY IN ORGANIC SOYBEAN. K. Crowley* and M. Ryan, Cornell University, Ithaca, NY

ABSTRACT

Cover crops can cycle nutrients, enhance soil health, and suppress weeds and therefore reduce the need for inputs and decrease environmental problems associated with nutrient losses, soil erosion, and non-target herbicide effects. It is important to consider how to maximize these biophysical benefits from cover crops, but farmers also need consider profitability. This research aimed to compare three cereal rye (*Secale cereale* L.) management strategies for their agronomic, soil health, and economic benefits in organic soybean (*Glycine max* (L.) Merr) production. In 2014-2015 and again in 2015-2016 in central New York, we compared 1) a No-till treatment, in which cereal rye was roller-crimped, followed by no-till planting soybeans, 2) a Ryelage treatment, in which cereal rye was harvested at boot stage for forage, followed by moldboard plowing, 3) a Plow Down treatment, in which cereal rye was terminated with a moldboard plow at jointing phase, and 4) a No Cover treatment, in which no cover crop was seeded and the field was moldboard plowed. Results show that weed biomass was greater in the No-till treatment than in the other three treatments in both years. No significant differences were observed in yield in 2015. In 2016, yields were lower in the No-till treatment, which was probably due to extremely dry conditions in June. Soil testing showed significantly better infiltration and higher soil respiration in the No-till treatment compared to the No Cover treatment, and significantly higher potentially mineralizable nitrogen in the Plow Down treatment compared to the No Cover and Ryelage treatments. No significant differences were observed in aggregate stability or active carbon. Preliminary results from a partial budget analysis show that harvesting cereal rye for ryelage and using tillage prior to growing organic soybean can maximize profitability.

EFFECTS OF CORN PLANTING DENSITY ON INTERSEEDED COVER CROPS AND WEED SEED PREDATION. C.Z. Youngerman*, A. DiTommaso, J. Losey, W. Curran, S.B. Mirsky, and M. Ryan, Cornell University, Ithaca, NY

ABSTRACT

Weed seed predation by invertebrates can be influenced by habitat quality. We established an experiment in NY, MD and PA in 2016 to evaluate the effects of corn planting density on the performance of interseeded cover crops, weed suppression, and weed seed predation in organic production. Corn was planted at three rates ranging from 25k-99k ha^{-1} and interseeded with a cover crop polyculture consisting of cereal rye, annual ryegrass, hairy vetch, and red clover at 66 kg seed ha^{-1}. Corn density treatments were arranged in a randomized complete block design, and included a control treatment with no corn and a control treatment planted at medium density with no interseeded cover crops. Cover crop and weed biomass were sampled at 50 and 110 days after interseeding. Invertebrate activity and diversity, and weed seed predation were measured at 50, 80, 110 and 140 days after interseeding. Weed seed predation was quantified using 25 seeds of *Setaria faberi* glued to seed arenas with and without invertebrate exclosures. Preliminary results show that average cover crop biomass at 110 days after interseeding ranged from 0.35 to 181.41 g m^{-2} in NY, and 0.13 to 62.43 g m^{-2} in MD, and was negatively correlated with corn density. Weed seed predation differed by sites and sample dates, and was lower on arenas with exclosures. A negative relationship between cover crop biomass and weed seed predation was observed in each of the three sites.

OPTIMIZING HERBICIDE AND COVER CROP PROGRAMS FOR INTEGRATED WEED MANAGEMENT IN NO-TILL GRAIN SYSTEMS. J.M. Bunchek*, J.M. Wallace, M. VanGessel, W. Curran, and D.A. Mortensen, The Pennsylvania State University, State College, PA

ABSTRACT

As the rise in herbicide resistance increasingly challenges successful weed control, adopting integrated weed management (IWM) becomes ever more important. IWM implements a multi-tactic approach to weed management that effectively controls pests and removes selection pressure away from herbicides. Tillage is often a mechanical control method utilized in an IWM program, but conservation tillage is a growing practice in the Mid-Atlantic due to the Chesapeake Bay Watershed Initiative. Thus, cover cropping and optimizing herbicide inputs are two IWM tactics of interest when continuous no-till is part of the rotation. To test this interest, we established a field experiment at the Russell E. Larson Agricultural Research Center (PSU-RELARC) near State College, PA, and the Carvel Research & Education Center (UD-CRED) near Georgetown, DE. This experiment aimed to quantify the efficacy of cover crop and herbicide programs on weed control and cash crop yield. We hypothesize that integrating cover crops with targeted herbicides will increase overall weed control efficacy. Further, this increased efficacy will reduce the need for herbicide inputs. The experiment used a randomized complete block split-plot design, with cover crop treatment as the main plot and herbicide program as the split. Cover crop treatments of cereal rye (135 kg ha^{-1}), a mixture of cereal rye and hairy vetch (100 kg ha^{-1} + 22 kg ha^{-1}), and a no cover control were drilled following corn silage harvest (Oct. 1-15). Cover crops were terminated with a burndown mixture ten days prior to planting glufosinate-resistant soybean. The four herbicide program splits were as follows: (1) a control program of burndown only (1.26 kg ae ha^{-1} glyphosate + 0.56 kg ae ha^{-1} 2,4-D LVE), (2) burndown + PRE-residual (1.7 kg ha^{-1} S-metolachlor + 56 g ai ha^{-1} flumetsulam), (3) burndown + POST (0.74 kg ai ha^{-1} glufosinate), and (4) burndown + PRE-residual + POST. Each split plot contained spiked 0.5 m2 microplots of smooth pigweed (*Amaranthus hybridus* L.) and horseweed (*Conyza Canadensis* [L.] Cronquist). The experiment in Delaware also included common ragweed (*Ambrosia artemisiifolia* L.) microplots. Weed seeds were collected from local populations prior to establishing the microplots, and both locations had natural populations of the weed species of interest. Each microplot was analyzed for weed density and size before and 10 days after treatment for each herbicide application. Additional data included late summer weed biomass (late July-early August) and soybean yield. The PSU-RELARC results showed that the addition of the PRE-residual at burndown decreased horseweed density and size; however, smooth pigweed populations were not controlled by the burndown application due to its emergence later in the growing season. Further, the POST glufosinate application was not necessary for horseweed control, since most weeds were effectively controlled with the burndown application. In addition, the POST application following the burndown + PRE-residual did not provide effective control of smooth pigweed as well as the burndown + POST. While smooth pigweed growth stage was reduced in programs that included the PRE-residual, the POST application was most effective in reducing smooth pigweed density. Integrating cover crops, especially cereal rye, can reduce weed biomass and help sustain soybean yield. Additionally, choosing an effective burndown application plus a PRE-residual or POST application is crucial for controlling problematic weeds while minimizing the herbicide resistance selection pressure.

CULTURAL WEED MANAGEMENT IN ORGANIC NO-TILL PLANTED SOYBEAN: A MULTI-TACTIC APPROACH. J. Liebert*, A. DiTommaso, and M. Ryan, Cornell University, Ithaca, NY

ABSTRACT

High soybean [*Glycine max* (L.) Merr.] populations have been shown to hasten canopy closure, which can improve both weed suppression and soybean yield. In conventional soybean production, the high cost of genetically engineered seed and seed treatments have led growers to plant at lower rates to maximize profitability. For organic farmers, market price premiums are typically double the price received for conventional soybean. In the absence of chemical and mechanical weed management, cultural practices are particularly important for adequate weed suppression in cover crop-based organic no-till planted soybean production. In 2014, an experiment was conducted in Aurora and Hurley, New York, to assess the effects of increasing soybean planting rates on weed suppression, soybean yield, and partial economic return. Planting rates of 198,000; 395,000; 595,000; 790,000; and 990,000 seeds ha^{-1} were arranged in a randomized complete block design. Weed biomass decreased and soybean yield increased at both sites as soybean population increased. The relationship between increasing soybean population and yield was described by an asymptotic model, and the predicted maximum yield was 2,506 kg ha^{-1} in Aurora and 3,282 kg ha^{-1} in Hurley. Lodging was negligible despite high soybean populations. Partial returns decreased beyond the predicted economically optimal planting rate of 650,000 seeds ha^{-1} in Aurora and 720,000 seeds ha^{-1} in Hurley as higher seed costs were no longer offset by yield gains. Based on our results, planting rates that are more than double the recommended rate of 321,000 seeds ha^{-1} for wide row ($\geq$76 cm) conventional soybean management in New York can enhance weed suppression, increase yield, and improve profitability in organic no-till planted soybean production.

EVALUATION OF MANAGEMENT OPTIONS FOR GLYPHOSATE, ALS- AND PPO-RESISTANT COMMON RAGWEED IN NORTH CAROLINA. B. Schrage* and W.J. Everman, North Carolina State University, Raleigh, NC

ABSTRACT

In 2016, a biotype of common ragweed (*Ambrosia artemisiifoia*) in northeastern North Carolina was confirmed to be resistant to glyphosate, ALS-, and PPO-Inhibiting herbicides. This discovery serves as the first agronomic dicot exhibiting resistance to three modes of action in North Carolina and prompted an evaluation into possible management options for growers in Camden, Chowan, Currituck, Gates, Hertford, Pasquotank and Perquimans counties.

With confirmation imminent, field experiments were conducted in Rocky Mount and Moyock, North Carolina during the summer of 2016 in order to evaluate: PRE programs, Liberty Link systems, POST options, residual herbicides, tank mixtures, and plant sizes during application that may influence efficacy. Rocky Mount experiments were not subjected to overwhelming wildlife populations or high levels of PPO-resistance enabling early canopy and effective control to take place. Analyzed separately, existing PRE, POST and tankmixed products effectively controlled common ragweed in all nearly all circumstances. In Moyock, the effectiveness and longevity of PRE herbicides was inconsistent prompting the recommendation of POST herbicides. Control was improved when POST applications were delayed to 3 or 4 WAP and additionally improved when a residual product was included with glufosinate. The utilization of a PRE herbicide greatly improved control. Fomesafen POST was statistically outperformed by other POST herbicides. Glufosiante, 2,4-D, and dicamba POST were able to effectively control PPO-resistant common ragweed at 15 cm; whereas, mesotrione failed to suppress ragweed applied greater than 10 cm.

SOYBEAN TOLERANCE AND PALMER AMARANTH (*AMARANTHUS PALMERI*) CONTROL OVER INCREASING RATES OF METRIBUZIN. T.A. Reinhardt*, D. Copeland, and W.J. Everman, North Carolina State University, Raleigh, NC

ABSTRACT

Broadleaf weed control has become increasingly challenging in soybean production due to multiple-herbicide resistant weeds. Metribuzin has been reconsidered to provide a unique mode of action in soybean productions, but is known to have variable soybean injury response. This research was organized in an RCBD with factorial arrangement to analyze the effect metribuzin rate on soybean injury, herbicide efficacy on Palmer amaranth, and final soybean yield. Adding metribuzin in a herbicide program to control multiple resistant Palmer amaranth would be helpful to resistance management, but rate needs to provide adequate control without sacrificing yield.

CRITICAL PERIOD OF GRASS WEED CONTROL IN SORGHUM. M.T. Schroeder* and W.J. Everman, North Carolina State University, Raleigh, NC

ABSTRACT

In North Carolina grass weed control in sorghum continues to be an issue. This is due to the limited POST herbicide options for grass control. To create an effective weed management plan for sorghum both the critical weed-free period as well as the critical time of weed removal need to be assessed. Two field studies were conducted near Clayton and Rocky Mount, NC to determine the effects of grass weed species on sorghum yield by allowing grasses to grow and compete with sorghum for various intervals of time. Hand weeding as well as herbicides applied to terminate the growth of grass species. This study is focused strictly on grass weeds so all broadleaf weeds were removed.

There were twelve treatments included in this study. 1) A pre-emergence only application was applied of atrazine and s-metolachlor 2&3) Weed free all season by hand removal as well as an added treatment where a herbicide was used 4) weedy all season 5-8) these treatments were left weedy until weeks 2,3,5, and 7, respectively and then weeds were removed for the entirety of the season 9-12) treatments were maintained weed-free until weeks 2,3,5, and 7, respectfully, and then weeds were allowed to grow throughout the entirety of the season. Grass weed species were then counted, collected, and documented at the end of each time interval. At harvest, biomass was taken from a meter of rows two and three of each treatment and both fresh and dry biomass was taken to document differences in growth from grass weed competition.

The PRE treatment showed differences in control depending on grass type that was present. Large seeded grasses, such as Texas Panicum (*Panicum Texanum*), that tend to emerge later in the season broke through the PRE application and effected the final biomass of the sorghum crop. These studies consistently showed stunting and reduction of final biomass in sorghum after week 5 and 7 of weed removal.

IMPACT OF COVER CROP QUALITY AND QUANTITY ON SUMMER ANNUAL WEED SUPPRESSION AFTER TERMINATION. K.B. Pittman* and M.L. Flessner, Virginia Tech, Blacksburg, VA

ABSTRACT

Summer annual weeds reduce crop yield through competition. Terminated cover crop residue can create a mulch layer that has the ability to suppress summer annual weed emergence. Research was conducted to evaluate summer annual weed suppression from cover crop monocultures and mixes and to correlate cover crop biomass and quality metrics, such as carbon to nitrogen ratios, to the duration of summer annual weed suppression.

This study was conducted in Blacksburg, Virginia as part of a larger experiment that is set up as a split-split plot design with 4 replications. Treatments consisted of four cover crops: cereal rye (*Secale cereale* L.) variety Elbon South, forage radish (*Raphanus sativus* L.) variety Nitroradish, hairy vetch (*Vicia villosa* Roth.) variety TNT, and crimson clover (*Trifolium incarnatum* L.) variety not stated, in monocultures and mixtures. Monoculture treatment rates were cereal rye at 125.4 kg ha^{-1}, hairy vetch at 28.0 kg ha^{-1}, crimson clover at 22.4 kg ha^{-1}, and forage radish at 9.0 kg ha^{-1}. Three two-way mixtures were included in this experiment, cereal rye at 50.4 kg ha^{-1} + hairy vetch at 20.2 kg ha^{-1}, cereal rye at 50.4 kg ha^{-1} + crimson clover at 15.7 kg ha^{-1}, and cereal rye at 69.4 kg ha^{-1} + forage radish at 4.5 kg ha^{-1}. Three-way cover crop mixtures contained cereal rye at 38.1 kg ha^{-1} and forage radish at 2.2 kg ha^{-1}, each with one legume: hairy vetch at 16.8 kg ha^{-1} or crimson clover at 13.4 kg ha^{-1}. A no cover crop control was also included. Cover crops were drilled on October 22, 2015 and terminated on April 26, 2016 using a roller crimper and glyphosate (Roundup PowerMAX®) at 1.26 kg ae ha^{-1}. At termination, above ground biomass was taken from a 0.09m^2 area. These samples were dried and sent off for analysis of carbon and nitrogen content (Feed and Water Analysis Lab; University of Georgia). Quality metrics of the cover crop mixture treatments were determined by using the mass determination from the biomass samples to determine a weighted average of the components. Visible weed suppression ratings were taken between rows on a two-week basis, starting 4 weeks after termination, using a 0 (no suppression) to 100 (complete suppression) scale as compared to the no cover crop check. Data were analyzed using JMP 11 and were subjected to ANOVA and effects were considered significant when $p < 0.05$, followed by a means comparison using Fisher's Protected LSD ($p < 0.05$) and correlations using logistic modeling.

Cover crop treatments containing cereal rye, either in monoculture or as part of a mixture, accumulated the most biomass (>8,400 kg ha^{-1}) followed by crimson clover and hairy vetch alone with 4,782 kg ha^{-1} and 2,813 kg ha^{-1}, respectively. This trend is similar with the C:N ratio. Cereal rye containing treatments had C:N ratios ≥22 while crimson clover alone and hairy vetch alone had C:N ratios of 11 and 9, respectively. Models combining rating data at 6 weeks after termination from pitted morningglory (*Ipomoea lacunosa* L.), redroot pigweed (*Amaranthus retroflexus* L.), large crabgrass (*Digitaria sanguinalis* L. Scop.), and carpetweed (*Mollugo verticillata* L.), predict 90% weed control from cover crops that produce >9,800 kg ha^{-1} biomass or have a C:N ratio above 21. Further research is needed to refine biomass and C:N ratios needed to suppress summer annual weeds and to determine the impact of high residue cover crops on cash crop yield.

SOYBEAN VARIETAL RESPONSE TO DICAMBA APPLIED AT THE VEGETATIVE GROWTH STAGE. A.M. Growe* and W.J. Everman, NCSU Crop Science, Raleigh, NC

ABSTRACT

With the spread of glyphosate resistant weed species throughout North Carolina, there has been a renewed interest of using growth regulator herbicides for weed control options in the state. As dicamba, a common growth regulator herbicide, is being incorporated back into herbicide programs, there is concern of off target movement to sensitive crops in adjacent fields. In 2015, 1.79 million soybean acres were harvested in North Carolina with a monetary value of 486 million dollars. Since the 1970's research has demonstrated that soybeans are highly sensitive to dicamba. As a major row crop, we must understand dicamba's affects on soybean cultivars grown in North Carolina. It is important to make dicamba applications during the best possible conditions so that risk of drift and volatility is minimized. To date, there has been little information reported on soybean varietal responses to dicamba drift.

The objective of these studies was to evaluate the effects of sub-lethal rates of dicamba on various group V and VI soybean cultivars at vegetative growth stages. Effects of dicamba were determined by collecting visual injury ratings, height reductions and yield reductions. Experiments were conducted at the Upper Coastal Research Station (Rocky Mount, NC), Peanut Belt Research Station (Lewiston, NC) and Lower Coastal Research Station (Kinston, NC) during 2015 and 2016. Five maturity group V and five maturity group VI soybean varieties were treated with dicamba at 1.1, 2.2, 4.4, 8.8, 17.5, 35, and 70 g ae ha^{-1} (1/512 to 1/8 of the labeled use rate for weed control in corn) during the V4 growth stage. Experiments were conducted using a factorial arrangement of treatments in a randomized complete block design, with two factors being soybean variety and dicamba rate. All data were subjected to analysis of variance and means were separated using Fisher's Protected LSD at $p= 0.05$.

A wide range of visual injury was recorded at 1, 2 and 4 WAT for all varieties. Higher levels of injury were associated with increasing dicamba rates. Injury ratings ranged from 25-73% and 19-66% 2 WAT for Maturity Group V and VI varieties respectively. Significant differences in height reductions to the non-treated check were also observed. Plant height reductions ranged from 22-72% and 11-57% 4 WAT for Maturity Group V and VI varieties respectively. Although there were no variety effects on height reductions, injury was greater for SS6810NR2, a maturity group VI variety. Although yield data has not been analyzed, soybean cultivars may be impacted by the effects of sub-lethal doses of dicamba. These studies demonstrate soybean's sensitivity to dicamba. It is important to make dicamba applications during the best possible conditions so that risk of drift and volatility is minimized.

COVER CROP AND WEED MANAGEMENT IN A LIVING MULCH SYSTEM USING REDUCED RATES OF HERBICIDES. V. Bhaskar*, R. Bellinder, A. DiTommaso, and S. Reiners, Cornell University, Ithaca, NY

ABSTRACT

After many decades of cover crop research and outreach, the use of cover crops to improve soil health and weed control continues to increase in many cropping systems. To further improve on the benefits of cover crops, their use as living mulches needs to be assessed. Living mulch systems provide a crucial enhancement of resource use as well as much needed soil protection from heavy rains and inter-row tillage early in the season. While many agricultural systems can benefit from inclusion of living mulches, they are especially suitable for small-scale, diversified farms. The main objective of this study is to assess whether a combination of living mulches and herbicide applications at reduced rates can decrease living mulch-cash crop competition and improve the reliability of weed control by living mulches. Because of the widespread use of herbicides in cropping systems, living mulches and their numerous benefits can be integrated in these systems as a way to advance agro-ecosystem sustainability. Herbicide applications are expected to enhance weed control while simultaneously reducing the vigor of the living mulch. The contribution of the living mulch towards weed suppression is expected to make the reduction in herbicide input feasible. In such a living mulch system, the crop, herbicide type and rate, and stage of growth of the living mulch and weeds at the time of herbicide application have to be carefully considered. Field trials were performed in 2015 and 2016 in Freeville, NY, using sunnhemp (*Crotalaria juncea* L.) as the cover crop species. While sunnhemp was established as living mulch in a tomato crop in a separate experiment, the results reported here were based on field trials using sunnhemp grown in pure stands only, to better assess interactions between the sunnhemp cover crop, weeds, and herbicides tested. The trials included twelve treatments set up in randomized complete block design with three replicates. Each plot was 1.83 by 3.1 m, with nine rows of sunnhemp, spaced 23 cm apart. Ten out of the 12 treatments consisted of two herbicide applications, each with a different type of herbicide. The herbicides evaluated and used in different combinations included fomesafen, halosulfuron, imazethapyr, metribuzin, rimsulfuron, and s-metolachlor. The remaining two treatments consisted of an untreated cover crop check with no herbicide application, and a weedy check with no cover crop or herbicide application. Cover crop and weed ground cover, and cover crop height were measured multiple times during the two growing seasons. Cover crop and weed density and aboveground biomass were determined at the end of each season. Because height of the living mulch would likely affect light availability for the cash crop, it was desirable that herbicides be capable of decreasing the height of the cover crop while maintaining considerable soil cover for effective weed control and soil protection. Reductions in cover crop height varied greatly depending on the herbicide used. However, soil cover by the sunnhemp was similar for all herbicide treatments, except the fomesafen + metribuzin treatment, where it was significantly lower. Weed biomass in weedy plots was 3.2 tons ha^{-1}, compared with 0.24 (92% reduction) to 1.0 (68% reduction) tons ha^{-1} in the cover crop-herbicide treatments. Similarly, weed cover in weedy plots was almost 95% compared with 4% to 45% in the cover crop-herbicide treatments. In treatments where no herbicides were applied, the sunnhemp cover crop reduced both weed biomass and weed cover by approximately 80%. The recovery in growth of the cover crop following the various herbicide application treatments was acceptable, except for the fomesafen treatment. Overall, herbicide applications reduced cover crop biomass by 0.76 to 3.08 tons ha^{-1}, height by 9 to 41 cm, and ground cover, by 1 to 57%. The results also indicated that a first herbicide application with strong post-emergence activity tended to significantly decrease cover crop biomass and cover, but with no corresponding decreases in weed biomass or cover. On the contrary, substantial injury to the sunnhemp cover crop led to increased weed pressure in these plots. Our experiments demonstrate that a combination of cover crops and herbicides at reduced rates is capable of providing adequate weed control, thereby making reduction in herbicide input feasible. Further research is required to develop strategies to increase control of cover crop height without decreasing soil cover.

THE IMPACT OF RESIDUAL HERBICIDES ON SOYBEAN VIGOR AND CONTROL OF MULTIPLE-RESISTANT PALMER AMARANTH. D. Copeland* and W.J. Everman, North Carolina State University, Raleigh, NC

ABSTRACT

Palmer amaranth (*Amaranthus palmeri*) has increasingly become one of the most problematic weed in the United States. Due to the rapid evolution of resistance of Palmer amaranth to multiple modes of action postemegence (POST), soybean growers ultimately rely on preemergence (PRE) herbicides for soil-residual activity. The objective of this study was to quantify the residual activity of different PRE herbicides and their effect on soybean vigor.

The effectiveness of 20 PRE herbicides was evaluated on Palmer amaranth in 2016 at locations in Cheraw, South Carolina and Whitakers, North Carolina. Herbicide treatments included: metribuzin (420 g ai ha^{-1}), linuron (560 g ai ha^{-1}), *S*-metolachlor (920 g ai ha^{-1}) + metribuzin (220 g ai ha^{-1}), pyroxasulfone (120 g ai ha^{-1}), *S*-metolachlor (1070 g ai ha^{-1}), dimethenamid-P (840 g ai ha^{-1}) , flumioxazin (72 g ai ha^{-1}), fomesafen (280 g ai ha^{-1}), fomesafen (266 g ai ha^{-1}) + *S*-metolachlor (1216 g ai ha^{-1}) , sulfentrazone (217 g ai ha^{-1}) + chlorimuron (28 g ai ha^{-1}), sulfentrazone (224 g ai ha^{-1}) + metribuzin(336 g ai ha^{-1}), sulfentrazone (122 g ai ha^{-1}) + *S*-metolachlor(1103 g ai ha^{-1}), sulfentrazone(260 g ai ha^{-1}) + cloransulam-methyl(34 g ai ha^{-1}), pendimethalin (800 g ai ha^{-1}), acetochlor (1260 g ai ha^{-1}), sulfentrazone(175 g ai ha^{-1}), flumioxazin (53 g ai ha^{-1}) + pyroxasulfone(68 g ai ha^{-1}), imazaquin (140 g ai ha^{-1}), dicamba (560 g ae ha^{-1}), isoxaflutole (53 g ai ha^{-1}) and a non-treated check was included. The experimental design was a RCBD design with four replications.

At both locations 2 and 4 weeks after planting (WAP), soybean dry weights per meter row resulted in less biomass when dicamba, isoxaflutole and flumioxazin + pyroxasulfone were applied. Palmer amaranth counts per square meter were greatest when dicamba, pendimethalin and imazaquin were applied. However, treatments including two, effective modes of action: sulfentrazone + chlorimuron, fomesafen + *S*-metolachlor, sulfentrazone + metribuzin and sulfentrazone + *S*-metolachlor resulted in less Palmer amaranth per square meter than all other treatments. This data demonstrates that PRE herbicides can delay early-season soybean growth, however, most treatments resulted in similar soybean biomass by 6 WAP. Also, growers should consider applying PRE herbicides that contain two modes of action for adequate Palmer amaranth control to avoid further delays of early-season soybean growth.

THE USE OF NATIVE WEEDS IN ECOLOGICAL RESTORATION. E.P. Law* and S. Diemont, Cornell University, Ithaca, NY

ABSTRACT

It's often said that a weed is a plant out of place, implying that plants we often struggle to control in agricultural systems can be beneficial in other contexts. One field where the use of native weedy plants shows great potential is ecological restoration. Restoration projects almost always involve revegetation of disturbed or degraded sites to restore structure and function to the ecosystem. Native species are preferred for revegetation due to their assumed pre-adaptation to climatic conditions, reduced risk of invasibility, and ecosystem services that support native fauna and thus enhance ecosystem functioning. The hardiness, reproductive capacity, and adaptations to disturbance that make weedy plants a nuisance to farmers also makes them ideal candidates to be used in these revegetation projects. In particular, these traits allow reassembled plant communities including native weedy species to better resist invasion by exotic species, a common problem in restored ecosystems. In many situations weedy species can be incorporated into restoration plans based on functional traits of the species that support restoration goals. In other cases, however, the restoration of particular native weeds that perform important ecosystem functions may be necessary due to a variety of factors that may exclude them from the landscape, causing restoration of said species to be a goal in and of itself.

In this presentation we will discuss the potential applications of native weedy plant species in ecological restoration using both basic concepts of ecological theory and two case studies from our recent research. The first case study involves the eco-cultural restoration of old fields in the northeast United States using the native edible weeds Jerusalem artichoke (*Helianthus tuberosus*), wild sunflower (*Helianthus annuus*), American groundnut (*Apios americana*), and common evening primrose (*Oenothera biennis*), in which provisioning and cultural ecosystem services were incorporated into restoration goals in addition to support of intended ecological outcomes. In this ongoing study the differential effects on reintroduced species and the surrounding plant community caused by mowing, tilling, and prescribed burning as site preparation are examined. The second case study considers the ecological interactions of the common milkweed (*Asclepias syriaca*), which is in rapid decline in the United States due to widespread use of glyphosate in GMO corn and soybean production systems. Milkweed's function as an important source of food for many insect species, including its role as the obligate host for Monarch butterfly larvae (*Danaus plexippus*), has led to many efforts to restore milkweed populations to preserve native insect habitat. These are just two examples of situations where native weedy species have been targeted for restoration in particular ecosystems where they provide services that enhance ecosystem function. Many other species that we often consider to be weeds may contain similar potential that begs to be explored.

TOLERANCE OF SEEDLING FORAGE GRASSES TO HERBICIDES. M.L. Flessner, S.C. Beam*, K.B. Pittman, S.C. Haring, and R.S. Randhawa, Virginia Tech, Blacksburg, VA

ABSTRACT

Weed control is critical for the success of forage grass establishment. However, little information is available regarding forage grass seedling tolerance to herbicides. A factorial field study was conducted in 2015 in Blacksburg, VA to determine the tolerance of common forage grasses to common pasture herbicides applied at various times after planting. Factors were grass species (perennial ryegrass (*Lolium perenne* L.), annual ryegrass (*Lolium perenne* L. subsp. *multiflorum* (Lam.) Husnot), Timothy (*Phleum pretense* L.), and orchardgrass (*Dactylis glomerata* L.)), herbicide treatment (aminopyralid + metsulfuron (Chaparral™) at 108 g ai + 16.5 g ai ha^{-1}, aminopyralid + 2,4-D (GrazonNext® HL) at 86 g ae + 699 g ae ha^{-1}, and triclopyr + fluroxypyr (Pasturegard® HL) at 630 g ae + 210 g ae ha^{-1}), and application timing (October 7, 14, 22, 29, and November 5 2015 which corresponded to 14, 21, 28, 35, and 42 DAP)). All species were drilled at 19 kg ha^{-1} on September 23, 2015. Treatments were arranged in a randomized complete split block design with a minimum of 3 replications per experimental unit. Treatments were applied with a handheld spray boom calibrated to deliver 140 L ha^{-1}. Visible ratings for cover reduction were recorded 2, 4, and 6 wk after treatment (WAT) and vigor 2 and 4 WAT. Cover reduction ratings ranged from 0 (no reduction) to 100 (stand loss), and vigor ratings ranged from 1 (no growth) to 10 (healthy stand). Data were analyzed in JMP 12 Pro using ANOVA with means separated by Fisher's protected LSD. Across all species aminopyralid + metsulfuron cover reduction was >80% and vigor was <3.8 4 WAT at all applications made on or before 35 DAP. Cover reduction was >60% and vigor was <5.5 4 WAT when applied 42 DAP. Aminopyralid + 2,4-D cover reduction was <43% and vigor >8 4 WAT when applied 21 DAP or later on perennial ryegrass, Timothy, and annual ryegrass. However, orchardgrass was more sensitive to aminopyralid + 2,4-D than any other species with >50% cover reduction 4 WAT at all application timings. Triclopyr + fluroxpyr had the least cover reduction of any herbicide treatment, <32% and vigor >7.8 4 WAT when applied sooner than 28 DAP, later applications had <20% cover reduction and vigor >9.3. Orchardgrass was more injured than the other species when triclopyr + fluroxypyr were applied sooner than 28 DAP. Overall forage grass seedlings have variable tolerance to the herbicides tested but in general, species tested were tolerant to herbicide apart from metsulfuron applied sooner than 42 DAP. Orchardgrass is more sensitive to these herbicides than the other species tested. Further research is needed to corroborate these findings and to determine the tolerance of these species to lesser herbicide rates.

HERBICIDE PROGRAMS FOR CRABGRASS AND GOOSEGRASS CONTROL ON CREEPING BENTGRASS GREENS. J.R. Brewer*, J. Craft, S. Rana, and S. Askew, Virginia Tech, Blacksburg, VA

ABSTRACT

Smooth crabgrass (*Digitaria ischaemum*) and goosegrass (*Eleusine indica*) are two of the most troublesome weeds on creeping bentgrass (*Agrostis stolonifera*) golf greens. There are currently no herbicides registered for postemergence crabgrass or goosegrass control on creeping bentgrass greens. Some research studies have indicated that repeated treatments of fenoxaprop at weekly intervals can selectively control crabgrass and goosegrass on creeping bentgrass greens. Since fenoxaprop can injure creeping bentgrass and is not registered for use on greens, there is need for research to evaluate herbicide programs for crabgrass and goosegrass control with minimal injury to creeping bentgrass greens. At Virginia Tech, we conducted three field trials during the summer of 2016 and two greenhouse studies during the fall of 2016. For field trials, putting greens were maintained at 3.2 mm height by daily mowing and consisted of '007' and 'L93' creeping bentgrass varieties at respective sites. Crabgrass and goosegrass control were evaluated on a USGA-specification green that was maintained fallow without creeping bentgrass. The greenhouse studies used 10 cm plugs of '007' creeping bentgrass placed in two separate greenhouses at the Glade Road Research Center.

Fenoxaprop and quinclorac were the most injurious treatment programs maintaining unacceptable injury (greater than 30%) for majority of the trial period. Fenoxaprop at 35 g ai ha^{-1} (biweekly) injured creeping bentgrass between 60 and 70% at both sites and more than other treatments, while quinclorac at 210 g ai ha^{-1} (biweekly) injured creeping bentgrass between 50 and 60% depending on site. Siduron applied at 5.6 kg ai ha^{-1} weekly or 13.5 kg ai ha^{-1} biweekly and topramezone applied at 1.5 g ai ha^{-1} weekly or 3 g ai ha^{-1} biweekly injured creeping bentgrass the least. Siduron injured creeping bentgrass 10% or less at both trial sites, while topramezone never injured creeping bentgrass more than 20%. The injury seen by topramezone only occurred for a short time after each of up to eight weekly treatments, and after some treatments no injury was observed. NDVI trends were similar to the observed injury. The greenhouse studies are also following the same visual injury trends as the field studies but are not yet complete. All treatments except the topramezone programs completely controlled crabgrass by week 7 on the 'L-93' creeping bentgrass green. Topramezone programs controlled crabgrass approximately 89% or better by week 7. Fenoxaprop completely controlled goosegrass and crabgrass at the fallow site, while quinclorac only controlled crabgrass. Siduron controlled crabgrass as well as fenoxaprop, and goosegrass slightly less. For example, siduron controlled goosegrass 80 and 94% when applied weekly or biweekly, respectively at 36 DAT and 43 and 67% when applied weekly or biweekly, respectively at 79 DAT. Topramezone controlled goosegrass as well as fenoxaprop, but was not as effective at controlling crabgrass. Results indicate that both topramezone and siduron can control crabgrass and goosegrass at commercially-acceptable levels with less injury than fenoxaprop or quinclorac.

TEMPORAL PATTERNS OF SEED SHATTERING AND GERMINABILITY OF COMMON LAMBSQUARTERS (*CHENOPODIUM ALBUM*) AND COMMON RAGWEED (*AMBROSIA ARTEMISIIFOLIA*). S.C. Haring*, M.L. Flessner, W.J. Everman, and S.B. Mirsky, Virginia Tech, Blacksburg, VA

ABSTRACT

Weed seed shattering phenology is an important biological trait for certain weed control technologies, such as harvest weed seed controls. Field experiments were conducted in 2015 in Blacksburg, Virginia to understand when seeds shatter from common lambsquarters (*Chenopodium album* L.) and common ragweed (*Ambrosia artemisiifolia* L.) in a soybean (*Glycine max* (L.) Merr.) field. Twenty-four individuals from each species were identified in a soybean field, and an 11.5 cm diameter sample cup was secured to the ground underneath each plant. Whole plants were monitored weekly for seed shattering, and weekly seed collections continued from the first sign of seed shattering until three weeks after the first possible soybean harvest, to simulate a harvest delay. Weekly seed collections were counted and then subjected to a direct germination assay in a growth chamber. Common ragweed began shattering seed four weeks before harvest time, and common lambsquarters began shattering three weeks before harvest. Both species had higher rates of seed shattering in their first two weeks of shattering, with common lambsquarters shattering 48% of total shattered seed in these two weeks, and common ragweed shattering 78% of its whole-experiment total. Seed shattering rates for both species then remained at a steady, lower level through the simulated harvest delay. After harvest, common lambsquarters and common ragweed each shed 40% and 13%, respectively, of their total shattered seed amounts. Total germination rates were 70% for common lambsquarters and 33% for common ragweed. No relationship was found between seed shattering date and germinability. These results suggest that seed dispersal occurs at varying rates across the growing season, peaking soon after weed maturity and continuing at meaningful levels through a delayed harvest. These results could have important implications for emerging harvest weed seed control technologies that rely on weed seeds to be retained at harvest time. Future research must be done to explore seed shattering phenology in other important weed species, as well as to relate these data to the total amount of seed produced by each plant and the total amount of seed retained through a harvest delay.

ZOYSIAGRASS RESPONSE TO HERBICIDES DURING GREENUP. J.M. Craft*, J.R. Brewer, and S. Askew, Virginia Tech, Blacksburg, VA

ABSTRACT

Turf managers in the transition zone have adopted the strategy of late winter applications of the non-selective and selective herbicides on bermudagrass (*Cynodon dactylon*) but are hesitant to apply late winter application to zoysiagrass (*Zoysia* ssp.) due to its perceived sensitivity to herbicides. Previous research suggest that proper timing can maximize weed control and ensure zoysiagrass safety. Three Virginia studies were conducted in the spring of 2016 to evaluate several herbicides and application timings based on growing degree day at base 32 F (GDD_{32}) accumulation for weed control and response of three different zoysiagrass varieties. At Westlake Golf Course near Hardy, VA, 18 treatments were evaluated for annual bluegrass (*Poa annua* L.) control and response of 'Zeon' zoysiagrass with herbicide treatments applied at 200 GDD_{32}. Two additional sites were conducted at Virginia Tech's Turfgrass Research Center in Blacksburg, VA, on a half and half mixture of 'Companion' and 'Zenith' zoysiagrass maintained at 5 cm height and 'Meyer' zoysiagrass maintained at 1.5 cm height with treatments applied at 400 and 600 GDD_{32}. At the 200 GDD_{32} application timing, no treatment caused more than 35% turf injury. At 4 WAT, glyphosate at 716 g ai ha^{-1}, glufosinate at 1628 g ai ha^{-1}, flumioxazin at 214 g ai ha^{-1}, foramsulfuron at 28 g ai ha^{-1}, flazasulfuron at 77 g ai ha^{-1}, and trifloxysulfuron-sodium at 16 g ai ha^{-1} controlled annual bluegrass 90% when applied at 200 GDD_{32}. Mixtures of flumioxazin and glyphosate controlled annual bluegrass less than either herbicide applied alone. At 4 WAT in lawn-height turf, glufosinate at 1682 g ai ha^{-1} applied at 400 or 600 GDD_{32}, glyphosate at 716 g ai ha^{-1} applied at 600 GDD_{32}, and rimsulfuron + metsulfuron methyl at 38.8 g ai ha^{-1} applied at 600 GDD_{32} reduced zoysiagrass greenup to less than 40% compared to more than 70% greenup from all other treatments. Although trends between herbicides were generally consistent between sites, the fairway-height 'Meyer' zoysiagrass was generally less injured than the lawn-height mixture. Finale at 1682 g ai ha^{-1} applied at 400 and 600 GDD_{32} 4 WAT had less green leaves per square decimeter than other treatments. Overall, glufosinate at 1628 g ai ha^{-1} was the most injurious treatment at all sites and all timings. Glyphosate did not injure zoysiagrass when applied at 200 or 400 GDD_{32} even though zoysiagrass green cover was 1 to 2.3% at the 400 GDD_{32} timing. All herbicides were safe to apply at 200 GDD_{32} and caused only transient zoysiagrass injury at most. At 400 GDD_{32}, glyphosate, diquat, oxadiazon, foramsulfuron, and rimsulfuron + metsulfuron injured zoysiagrass less than 30% but glufosinate and flumioxazin injured zoysiagrass more than 30% and more than other treatments. Treatments applied at 600 GDD_{32} were more prone to cause injury and delay spring greenup for most herbicides.

COMPARISON OF TWO GREEN MANURE COVER CROPS AND VARYING TILLAGE TIMING FOR MANAGING WEEDS IN AN ORGANIC CORN PRODUCTION SYSTEM. R. Champagne*, W. Curran, J.M. Wallace, and T.R. Mazzone, Penn State University, State College, PA

ABSTRACT

In organic agriculture, tillage is used to control weeds, prepare a seedbed, and incorporate nutrient amendments. However with increasing concern for soil health and water quality, many farmers are looking for ways to reduce tillage in their operations. Continuous no-till in organic row crop production is currently not feasible, as several agronomic problems can arise, mainly related to weed control, cover crop termination, and crop establishment issues. To research ways to reduce tillage in organic operations, a cropping systems study was initiated at Penn State University in fall 2014 to study alternative strategies for reducing tillage frequency and intensity within an annual organic grain rotation. Two green manure cover crops before corn (*Zea mays* L. 'MC4050') and varying tillage timing are being implemented in four systems (S1-S4) to help reduce tillage and control weeds. S1 and S2 compare hairy vetch (*Vicia villosa* 'Groff') plus triticale (*Triticosecale*) before corn grown for silage and grain. S1 involves fall tillage and manure application so that the corn may be no-till planted into the rolled cover crop in the spring. In-season weed management is achieved through use of a high-residue cultivator. S2 carries out standard broadcast manure application and use of a moldboard plow in the spring to terminate the hairy vetch cover crop. S3 and S4 compare red clover (*Trifolium pretense* L.) plus timothy (*Phleum pretense* L.) frost-seeded using a no-till drill into a winter cereal. Manure application and plowing are used to terminate the cover crop prior to rotation to corn silage or grain. S2-S4 achieves in-season weed management through standard inter-row cultivation. Cover crop biomass, weed biomass, and cash crop yield are being examined as indicators of the success of these alternative strategies to help reduce tillage in organic cropping systems. Preliminary results show that hairy vetch plus triticale produces significantly higher spring cover crop biomass than red clover plus timothy when terminated on the same date and approximately one to two weeks later- S1 and S2 together averaging around 5,900 kg ha^{-1} compared to about 3,700 kg ha^{-1} in S3 and S4. However, successful termination of hairy vetch plus triticale with no-till is more challenging in the S1 system with corn, yielding less than in the S3 red clover plus timothy system (averaging 24,600 vs. 38,100 kg ha^{-1}, respectively). A number of factors likely contributed to this difference, including later planting date and stand establishment issues in no-till and differences in N mineralization and weed control. In contrast, first year results suggest S2 and S4 produce similar corn grain yields (around 9,100 kg ha^{-1}) when the cover crops are terminated with tillage and corn is planted on the same date. Across all four systems, there were no differences in late summer weed biomass, and dry matter was generally less than 1,000 kg ha^{-1}. This project will continue through fall 2017 aiming to refine organic reduced tillage tactics in hopes of producing beneficial results for grain and silage farmers in the mid-Atlantic region.

A NEW *S*-METOLACHLOR PLUS DICAMBA PREMIX AS AN EFFECTIVE TOOL IN AN INTEGRATED WEED MANAGEMENT PROGRAM IN DICAMBA TOLERANT SOYBEANS. E. Hitchner*, A. Moses, D. Porter, T. Trower, and J. Holloway, Syngenta, 08098, NJ

ABSTRACT

A new low volatility premix formulation of *S*-metolachlor plus dicamba is under development by Syngenta Crop Protection for weed control in dicamba-tolerant soybeans and cotton. The dual site-of-action herbicide is designed to deliver pre- and post-emergence activity as a burndown, pre-emergence or early post-emergence application. Field trials were conducted in 2015 and 2016 to evaluate weed control efficacy and crop safety of this new herbicide as part of an integrated weed management program in dicamba-tolerant soybeans.

The *S*-metolachlor plus dicamba formulation provides post-emergence control of many important weed species including horseweed, common and giant ragweed and common lambsquarters. It also provides post-emergence and residual control of *Amaranthus* species and residual control of many annual grass weeds. Successful and consistent weed control in dicamba-tolerant soybeans are targeted with programs that include an effective burndown, pre-emergence residuals and post-emergence herbicides with multiple, overlapping sites of action. *S*-metolachlor plus dicamba will be an important herbicide tool as part of an integrated weed management program for dicamba-tolerant soybeans.

PHYSIOLOGICAL AND MOLECULAR ANALYSIS OF GLYPHOSATE RESISTANCE IN GIANT RAGWEED (*AMBROSIA TRIFIDA*). B. Schulz*, K. Segobye, and S.C. Weller, University of Maryland, College Park, MD

ABSTRACT

Giant ragweed *(Ambrosia trifida L.)* is a competitive annual weed prominent in agriculture fields and cultivated areas. It is one of the most problematic weeds in corn and soybean production systems. The introduction of glyphosate resistant agronomic crops ("Roundup®-ready") in 1996 provided an effective tool to manage giant ragweed. The use of glyphosate based herbicides drastically increased after 1996 due to overreliance and repeated use of glyphosate for weed control in glyphosate resistant crops. These use patterns resulted in very high selection pressure for glyphosate resistant giant ragweed (GRGR). GRGR has been reported in several states of the U.S and in Canada. The ultimate goal of our project is to discover and isolate genes responsible for glyphosate resistance in GRGR to design informed strategies for weed control of GRGR plants. An unusual mode of resistance to glyphosate was found in Giant ragweed. A rapid response of glyphosate treated leaves in GRGR, which show a hypersensitive-like (HR) reaction to herbicide treatments was observed. This HR results in leaf abscission within a day of treatment. GRGR plants do not die from glyphosate treatments but resume normal-growth from shoot and axillary meristems, which do not show an HR-like response and reproduce. Glyphosate sensitive plants do not show a comparable rapid HR and transport the herbicide systemically throughout the entire plant, which leads to leaf chlorosis and eventual death of the treated plants after 2-3 weeks, likely due to inhibition of EPSPS, the key enzyme of the shikimate pathway.

Glyphosate resistant and sensitive biotypes were compared for rate response to glyphosate at typical field rates of 1x to 8x application rates (1x=0.84kg ha^{-1}). We found that GRGR plants survive rates of up to 6.72 kg/ha.

Physiological analysis suggested that signaling pathways related to pathogen response systems of plants as well as basic metabolic regulation controls the response of GRGR plants to glyphosate treatment.

To assess the reaction of sensitive and resistant biotypes to glyphosate on the molecular level we analyzed the total transcriptome of treated and untreated plants after glyphosate application. We have identified a list of genes that are differentially expressed between the two biotypes as the first step in identifying genes responsible for the glyphosate resistance observed. Combined with genetic analysis we developed a strategy for gene identification of glyphosate resistance genes in GRGR plants.

GRAPE HYACINTH (*MUSCARI BOTRYOIDES*) CONTROL IN A WHEAT-SOYBEAN ROTATION.

M.L. Flessner* and M. VanGessel, Virginia Tech, Blacksburg, VA

ABSTRACT

Grape hyacinth [*Muscari botryoides* (L.) Mill] is a difficult-to-control, perennial weed that emerges in fall, coinciding with senescence and harvest of summer crops. The fall leaf production can interfere with harvesting the lower soybean [*Glycine max* (L.) Merr.] pods. Typical of the Hyacinthaceae, it produces numerous bulbs that allow it to spread. Little information is available on the response of grape hyacinth to herbicides. Research was conducted to evaluate several herbicides labeled for use in wheat or for preplant burndown. Fall applied and spring applied experiments were conducted, respectively. Both experiments were conducted in a randomized complete block design with 4 replications in a no-till field in Essex County, Virginia. Experiment one was applied 16 November 2015, and experiment two was applied 5 April 2016. Data collected included visible green cover and visible control relative to the nontreated check. Additionally, experiment one evaluated grape hyacinth density via line intersect method while experiment two evaluated grape hyacinth bulb density via sieved soil samples and subsequent dry biomass. Soybeans were grown across both experiments with fomesafen at 1.21 kg ha^{-1} + *S*-metolachlor at 0.27 kg ha^{-1} (Prefix) + glyphosate applied 2 May 2016 and a POST application of glyphosate in early June. Ground cover at the preplant application was less than 5% visible green grape hyacinth.

Results for experiment one indicate that paraquat (0.56 kg ha^{-1}) and paraquat + metsulfuron (8.4 g ha^{-1}) resulted in >90% control 3 weeks after treatment (WAT). Carfentrazone (35 g ha^{-1}) resulted in 80% control 3 WAT, while all other treatments resulted in ≤50% control. At 6 WAT glyphosate (1.26 kg ae ha^{-1}) + dicamba (0.56 kg ae ha^{-1}) and paraquat + metsulfuron provided the best control, 88 and 80% respectively. Paraquat alone resulted in significantly less control (51%), 6 WAT indicating recovery was occurring. From February to May, control from group 2 herbicides increased. Group 2 herbicides were in the top performing statistical grouping as rated 2 May 2016; metsulfuron at 8.4 and 16.8 g ha^{-1} resulted in 75 and 77% control, respectively, while sulfosulfuron (35 g ha^{-1}) resulted in 85% control and chlorsulfuron (17.5 g ha^{-1}) + metsulfuron (3.5 g ha^{-1}) resulted in 70% control. Glyphosate and glyphosate + dicamba resulted in 81 and 48% control, respectively, at this timing and were not different than group 2 herbicide treatments. No differences in visible cover or density were observed after soybean harvest on 1 November 2016 relative to the nontreated check. Results for experiment two indicate that paraquat containing treatments resulted in >90% control 4 WAT. Glyphosate alone resulted in 77% control while glyphosate + dicamba resulted in 83% control 4WAT; neither treatment was different than paraquat containing treatments. All other treatments resulted in <55% control 4 WAT. No differences in visible cover, bulb density, or bulb weight were observed after soybean harvest on 1 November 2016 relative to the nontreated check. Future research should investigate other techniques to reduce or eradicate grape hyacinth over time, such as sequential herbicide applications or tillage.

EFFICACY OF PENDIMETHALIN AND QUINCLORAC FOR CONTROL OF LARGE CRABGRASS (*DIGITARIA SANGUINALIS L.*) IN ORCHARDGRASS. Q. Johnson*, M. VanGessel, R. Chandran, and M.L. Flessner, University of Delaware, Georgetown, DE

ABSTRACT

Large crabgrass (*Digitaria sanguinalis* (L.) Scop.) and other annual grass weeds are problematic in cool-season grass forage due to competition, which can result in reduced forage or hay quantity and quality. Until recently, control options for annual grasses in perennial grass pastures or hayfields have been very limited. Research was conducted to evaluate pendimethalin (Prowl H20) and quinclorac (Facet L) for large crabgrass control. Two experiments were conducted in Lincoln, DE in randomized complete block designs with 3 replications. Experiment one evaluated pendimethalin at 1, 2, or 4 lb ai/A applied PRE to weeds (application A) or 2 lb ai/A applied PRE to weeds (application A) and after first cut (application B). Experiment two evaluated pendimethalin at 2 lb ai/A applied PRE to weeds (application A) fb 2 lb ai/A after first cut (application C); pendimethalin at 4 lb ai/A applied PRE to weeds (application A) fb quinclorac at 0.4 lb ae/A applied POST (application D); quinclorac at 0.4 lb ae/A applied POST on 1-2" crabgrass (application B); and quinclorac at 0.4 lb ae/A applied POST (application D). Quinclorac was applied with methylated seed oil at 1.25% v/v. Both experiments included non-treated running checks. Application A was 18 March 2016, applications B and C were 31 May 2016, and application D was 2 September 2016. Applications A and B were timely, application C was delayed slightly due to delayed first cutting, and application D was delayed significantly due to delayed second cutting. Percent stand was determined at the end of the season (6 October 2016) by counting component of stand (orchardgrass (*Dactylis glomerata* L.), large crabgrass, bare ground, other weeds) every foot along a 100-foot tape in the center of each plot. In experiment one, the length of residual large crabgrass control increased with increase in pendimethalin rate. The 4 lb ai/A rate provided approximately 12 weeks of control, with orchardgrass and large crabgrass ground covers at 48% and 28% respectively, while the split application provided season long control, with orchardgrass and large crabgrass ground covers at 69% and 3%, respectively. In experiment two, quinclorac POST (application B) provided excellent initial large crabgrass control, but residual control was limited to approximately 6 weeks, with orchardgrass and large crabgrass ground covers at 38% and 23%, respectively. Quinclorac at that timing also caused significant orchardgrass response (16% stunting) that lasted approximately 6 weeks. Sequential applications provided the best control of large crabgrass. Pendimethalin fb quinclorac (applications A and D) or pendimethalin fb pendimethalin (applications A and C) provided season-long large crabgrass control with orchardgrass covers of ≥58% and large crabgrass ground covers of <3%.

IN VITRO EFFECTS OF SORGOLEONE (SORGHUM ROOT EXUDATE) ON GROWTH OF DIFFERENT WHEAT AND WEED SPECIES. M.K. Bansal* and W.J. Everman, North Carolina State University, Raleigh, NC

ABSTRACT

Sorghum in known to produce the allelochemical sorgoleone. Production of sorgoleone is a dynamic process where plants produce these chemicals either by roots when they are still alive or by dead decaying matter and are known to have negative impact on weeds and following crops. There are concerns about sorghum affecting the following winter wheat crop when grown in rotation in North Carolina. Impact of sorgoleone on growth of wheat and various weed species was investigated under in-vitro conditions. Seeds of wheat (Shirley and USG3251) and four weed species, large crabgrass, Italian ryegrass, velvetleaf, and sicklepod were pre-germinated and then transferred to 20x100mm petri dishes treated with varying concentrations of sorgoleone. Sorgoleone was applied @ 0 (control), 25, 50, 100, 150, 200, and 300 $\mu g\ ml^{-1}$. 10 days after placing seeds on the petri dishes, growth was measured in terms of shoot length. Significant sorgoleone treatment effects were observed for shoot growth when pooled over species. Shoot length was reduced at higher rates of sorgoleone compared to control treatments for all weed species. Wheat shoot length was not significantly affected by sorgoleone concentration. Velvetleaf shoot length was lower at all concentration compared to the non-treated control. At higher rates of sorgoleone, large crabgrass, Italian ryegrass, and sicklepod growth was reduced when compared to lower rates. Results suggests that sorgoleone has a negative impact on growth of weed species, however the wheat cultivars tested were not impacted.

COVER CROP TRAITS AND MANAGEMENT STRATEGIES INFLUENCE HORSEWEED SUPPRESSION IN NO-TILL CROPPING SYSTEMS. J.M. Wallace*, W. Curran, M. VanGessel, D.A. Mortensen, J.M. Bunchek, and B. Scott, Penn State University, State College, PA

ABSTRACT

Multi-tactic weed control strategies are needed to improve management of glyphosate resistant horseweed (*Conyza canadensis* [L.] Cronquist) in conservation tillage systems within the Northeastern United States. Current management recommendations are to include multiple herbicide sites of action in burndown programs. Integration of fall-planted cover crops has the potential to improve burndown management of horseweed and move selection pressure away from effective burndown herbicides, but greater understanding of cover crop management practices are needed to optimize suppression of horseweed populations. Previous research has demonstrated that interactions among cover crop traits, growing season length and soil fertility can significantly impact the level winter annual weed suppression. We conducted field experiments to evaluate cover crop strategies for horseweed management in 2014-2015 and 2015-2016 at Penn State's Russell E. Larson Agricultural Research Center (PSU-RELARC) in central PA and at University of Delaware's Carvel Research and Education Center (UD-CREC) near Georgetown DE. The objectives of our research were to compare the effect of: 1) winter-hardy and winter-kill cover crops traits, 2) grass-only and grass-broadleaf mixtures, 3) low- and high-residual soil fertility, and 4) long- and short-fall growing season windows on horseweed populations at time of spring burndown applications. In the first experiment, we evaluated seven cover crop treatments for long growing season windows (early-Sept plant date): no cover, cereal rye (134 kg ha^{-1}), spring oats (134 kg ha^{-1}), cereal rye + hairy vetch (67 + 22 kg ha^{-1}), cereal rye + forage radish (67 + 6 kg ha^{-1}), spring oats + hairy vetch (67 + 22 kg ha^{-1}), and spring oats + forage radish (67 + 6 kg ha^{-1}). Split-plots were imposed to provide a low- and high-residual soil fertility gradient: 0 or 67 kg N ha^{-1} at planting using AMS. In the second experiment initiated in 2015-2016, we compared a subset of these cover crop treatments using long (early-Sept plant date) and short (early- to mid-Oct plant date) growing season windows. Glyphosate burndown applications were made prior to cover crop planting and 42 kg N ha^{-1} was applied prior to planting for long-season cover crops and as a spring topdress application for short-season cover crops. Prior to planting cover crops, locally-collected horseweed seed was distributed in permanently marked microplots (0.50 m^2) at an average rate of 5,400 seeds m^{-2}. Cover crop biomass (kg ha^{-1}) and horseweed density and size were collected 10 weeks after planting (WAP) and at spring burndown. Cover crops were terminated at the cereal rye boot stage (Zadok 45) using glyphosate + 2,4-D (1.26 + 0.56 kg ha^{-1}). The following results are for the Pennsylvania location only. In a long fall growing season window, our results showed that each cover crop treatment resulted in significantly lower horseweed population densities (57 to 96%) at the time of the spring burndown application. Reductions in horseweed populations were significantly greater in high-residual soil fertility plots (88%) compared to low-residual plots (65%) across cover crop treatments. The use of cereal rye, a winter-hardy grass, improved horseweed suppression compared to spring oats, a winter-kill grass, in high-fertility plots but not low-fertility plots. The use of cover crop mixtures did not improve horseweed suppression compared to oats or cereal rye alone, with the exception of oats + radish in high fertility plots, which resulted in greater horseweed suppression (92%) compared to oats alone (82%). Our results indicate that winter-hardy cover crops can result in significantly lower horseweed population densities and also reduce mean horseweed rosette diameters at the time of the spring burndown application in comparison to the control and winter-kill cover crop treatments. Finally, the use of cereal rye in the short-growing season window resulted in comparable levels of horseweed suppression to cereal rye in the long-growing season window. Our results suggest that cover cropping is a viable strategy for improving horseweed management, but residual soil fertility and functional traits of cover crops will determine the level of horseweed suppression.

EFFICACY OF QUINCLORAC AND PENDIMETHALIN FOR GIANT FOXTAIL (*SETARIA FABERI*) CONTROL IN HAYFIELDS. M.L. Flessner*, R.S. Randhawa, Q. Johnson, and R. Chandran, Virginia Tech, Blacksburg, VA

ABSTRACT

Annual foxtails (*Setaria* spp. P. Beauv.) are problematic grassy weeds in hayfields due to their bristly seedheads, which are capable of puncturing mucous membranes leading to injuries to mouth and gum areas, particularly in horses. Herbicidal control options for foxtails are limited and little information exists regarding their efficacy. Research was conducted to evaluate pendimethalin (Prowl H20; BASF) and quinclorac (Facet L; BASF) for giant foxtail (*Setaria faberi* Herrm.) control. Two experiments were conducted in Blacksburg, VA in randomized complete block designs with four replications per treatment. Experiment one evaluated pendimethalin applied PRE (application A) at 1.17, 2.24, 4.5 and 2.24 followed by (fb) 2.24 kg ha^{-1} 30 days apart (application B). Experiment two evaluated pendimethalin PRE (application A) at 4.5, pendimethalin PRE 2.24 fb 2.24 kg ha^{-1} 30 days apart (application A fb B), pendimethalin PRE at 4.5 kg ha^{-1} fb quinclorac at 420 g ae ha^{-1} at first hay cutting (application A fb C), quinclorac at 420 g ha^{-1} early POST (application B), and quinclorac at 420 g ha^{-1} at first cutting (application C). Quinclorac was applied with methylated seed oil at 1.67% v v^{-1}. Both experiments included a nontreated check. Application A was 13 April 2016, application B was 5 May 2016, and application C was 8 June 2016 for both experiments. No foxtail emergence was observed at initiation for both experiments. Visible control data (0 to 100% scale with 0 corresponding to no control and 100 corresponding to complete control relative to the nontreated check) were collected monthly after the PRE treatment for 4 total ratings for both trials. Results indicate that pendimethalin at ≥2.24 kg ha^{-1} maximized giant foxtail control, which was 73 to 90% until mid-July across experiments. Subsequently, these treatments declined to 65 to 88% control in experiment one and 55 to 60% control in experiment two. Quinclorac alone resulted in the better control when applied after the first cutting (85 to 90%) compared to early POST (55 to 81%), across June to August ratings. Pendimethalin fb quinclorac resulted in 90 to 100% control across all rating dates. This result was not different than quinclorac applied alone after the first cutting for all applicable rating dates. These studies indicate that giant foxtail can be controlled in hayfields with both quinclorac and pendimethalin. Future research is needed to assess efficacy on other problematic annual foxtail species.

COVER CROP TERMINATION TIMING CAN AFFECT WEED CONTROL AND CROP PERFORMANCE; THE PROS AND CONS OF PLANTING GREEN. W. Curran*, D. Lingenfelter, and H. Myer, Penn State University, University Park, PA

ABSTRACT

Traditionally cover crops are terminated 7 to 14 days before planting. "Planting green" is becoming more common with some innovative farmers and has been defined as the planting of the primary cash crop into an actively growing cover crop generally using a no-till planter or drill. Corn (*Zea mays* L.), soybean (*Glycine max* [L.] Merr.), and other large seeded crop species have the greatest likelihood of success using this technique. There are a number of perceived benefits to this practice including increased above and below-ground cover crop biomass, greater N-fixation from legumes such as hairy vetch or crimson clover because of delayed termination, and conservation of soil nutrients for future crops and less environmental nutrient loss. In addition there can be a reduction in excess soil moisture which can sometimes limit spring planting, more living roots and the potential for greater soil cover to decrease soil erosion, help in building or maintaining soil structure and nutrient cycling, and reduced compaction. Finally, with the cover crop mulch, there can be reduced soil moisture loss from evaporation, and the potential for reduced slug herbivory to the primary cash crop. Some potential disadvantages include the need for specialized no-till equipment for successful cash crop establishment as high residue systems can be more challenging. Certain insect pests could be more problematic in living cover crop systems and dry spring weather could lead to a soil moisture deficit impacting successful cash crop establishment and growth. Immobilization of soil nitrogen by more mature grass cover crops can necessitate the need for additional nitrogen at planting time and especially for corn. Finally, herbicide selection and application timing can be impacted by planting green since multiple herbicides may be needed to control the cover crop as well as any weeds. We have been investigating the planting green concept at Penn State University in replicated field trials comparing planting green to terminating the cover crop at a more traditional time. Our results thus far have shown that soybean performance is similar regardless of cover crop termination timing assuming an adequate soybean population is achieved. For corn, our results have been less consistent with lower yields in some studies when planted green compared to earlier termination of the cover crop. The reduced yield has been a result of reduced corn population and/or what we believe is insufficient N availability due to N immobilization by the cover crop. In some corn herbicide trials that included 2,4-D and dicamba in the burndown program, corn was not negatively impacted by the herbicide regardless of application timing. However, we recognize that both air temperature and timing of rainfall relative to herbicide application and corn planting can influence the potential for crop injury. Future studies should investigate alternative burndown herbicides and/or the potential fit for new herbicide resistant crop technologies to be used with this no-till cover crop management approach.

DEVELOPMENT OF AN EDUCATIONAL MAPPING TOOL FOR DOCUMENTING AND RESEARCHING THE SPREAD OF HERBICIDE-RESISTANT WEEDS IN THE US. A. Klodd*, W. Curran, D. Miller, S. Crawford, D. Lingenfelter, and A. Davis, Pennsylvania State University, University Park, PA

ABSTRACT

The spread of Palmer amaranth and waterhemp, two competitive herbicide resistant pigweeds, is becoming an increased challenge for row crop production in many US states. As multiple resistant biotypes of these weeds continue to spread geographically, awareness of their locations may help producers prepare appropriate prevention and management plans before infestations become severe. Additionally, learning the geographic distribution and dynamics of resistant pigweed populations could aid research efforts that aim to understand the causes and rate of spread. To gather information on the national distribution of Palmer amaranth and waterhemp resistance biotypes, an online mapping tool of all US states is currently being developed to crowdsource this data and make it easily viewable by stakeholders. The completed tool will enable voluntary data entry through an online form, including location (GPS or county), species, crop, and multiple management practices utilized on the affected area. The real-time public map displays the number of unique locations entered per county for each species, while all other data is accumulated in a non-public database for research use. Research data may be analyzed to correlate management variables to the spread of pigweed geographic range. The primary intended result of this tool is to create easy access to accurate and thorough distribution information to aid in decreasing the spread and development of herbicide resistant pigweed populations.

INTEGRATING TILLAGE, COVER CROPS, AND HERBICIDES FOR WEED MANAGEMENT IN DELAWARE. K.M. Vollmer*, M. VanGessel, Q. Johnson, and B. Scott, University of Delaware, Georgetown, DE

ABSTRACT

As herbicide-resistant weeds continue to be a problem in cropping systems, a need for more integrated forms of weed management are needed. Methods for integrated weed management include preventative, cultural, mechanical, biological, and chemical. In 2016, a study was conducted at the University of Delaware Carvel Research and Education Center to examine the effects of integrating mechanical, cultural, and chemical weed control systems in corn and soybean. The study was arranged as three-factor split plot with tillage type being the whole plot (fall moldboard plow, spring tillage, and no-till), cover crop (with fall seeded or without) being the split-plot, and PRE herbicide treatment (yes or no) being the split-split-plot. A mixture of cereal rye (34 kg ha^{-1}) + hairy vetch (34 kg ha^{-1}) was used as the cover crop prior to corn planting, and cereal rye alone (135 kg ha^{-1}) was used prior to soybean planting. Cover crop biomass and winter annual weed densities were taken prior to cover crop termination. In corn, atrazine (1.8 kg ha^{-1}) + *s*-metolachor (1.4 kg ha^{-1}) was applied at planting. In soybean, flumioxazin (0.07 kg ha^{-1}) + pyroxasufone (0.13 kg ha^{-1}) + paraquat (0.8 kg ha^{-1}) was applied at planting. Summer annual weed densities were taken 4 weeks after planting (WAP). Glufosinate (0.5 kg ha^{-1}) + dimethanamid-P (1 kg ha^{-1}) was applied 4 WAP as a broadcast application to control emerged weeds in both corn and soybeans. Data was analyzed for analysis of variance using the general linear model in JMP Pro 12 and means separated using Fisher's LSD ($\alpha = 0.05$)

In the corn plots, there were no differences in cereal rye biomass due to tillage system prior to corn planting but hairy vetch biomass was higher in plots that had been moldboard plowed compared others. Common chickweed, cutleaf evening primrose, and field pansy counts were lower in plots containing a cover crop or in plots that had been moldboard plowed. Common ragweed density was higher with spring tillage and no herbicide, while all other plots were comparable to one another. Palmer amaranth density was higher in no-till and no PRE herbicide treatment. Percent weedy grass cover was highest in no-till and no PRE herbicide treatment, followed by spring tillage and no PRE herbicide, while all other treatments had similar density of grasses. Corn height was lower in spring-tillage with no PRE herbicide. Corn yield was higher in moldboard plow plots compared to spring tillage.

Cereal rye biomass in moldboard plow treatments was higher compared to other tillage systems prior to soybean planting. Common ragweed density was higher in spring-tillage with no PRE herbicide, while all other treatments were comparable to one another. Palmer amaranth density was higher in spring tillage and no-till plots with no PRE herbicide. Percent weedy grass cover was higher in all spring-tillage plots and in no-till plots with no PRE herbicide. Soybean heights were higher in no-till compared to spring-tillage. Soybean yield was higher in spring-tillage plots compared to other treatments.

Cover crop establishment was much better with moldboard plowing. Low cover crop biomass in the spring is one reason there was little effect of cover crop on weed densities. This trial will be repeated, with corn plots rotating to the respective soybean systems and vice versa. This will allow for evaluation of multi-year impact of integrated weed management. This data shows that the presence of a cover crop has less of an effect on annual weed suppression compared to tillage system and the use of PRE herbicides. More research is needed to examine the effects of these treatments over successive years.

WHICH POST HERBICIDES WILL CONTROL MARESTAIL IN AGRONOMIC CROPS? D. Lingenfelter* and W. Curran, Penn State University, University Park, PA

ABSTRACT

A field study was conducted in 2016 in Pennsylvania (Landisville, Lancaster Co.) to evaluate glyphosate-resistant horseweed/marestail (*Conyza canadensis*) control with POST herbicides in a fallow setting. Studies were arranged in a randomized complete block design with three replications. Herbicides were applied with a small-plot, CO_2-backpack sprayer system that delivered 15 GPA thru TeeJet AIXR110015 nozzles on June 10 and marestail ranged from 4-14 inches tall (8 inch average height). Treatments included: glyphosate (1.13 lb ae/A), cloransulam (0.0315 lb ai), chlorimuron (0.0105 lb), imazamox (0.039 lb), fomesafen (0.25 lb), lactofen (0.195 lb), acifluorfen (0.375 lb), fluthiacet (0.0064 lb), saflufenacil (0.0223 lb), glufosinate (0.66 lb), bentazon (1 lb), dicamba (0.25), dicamba + diflufenzopyr premix (0.175 and 0.35 lb ae), 2,4-D ester (0.5 lb ae), fluroxypyr (0.14 lb), clopyralid + flumetsulam premix (0.196 lb), mesotrione (0.094 lb), topramezone (0.0164 lb), bromoxynil (0.375 lb), pyrasulfotole + bromoxynil premix (0.241 lb), halosulfuron (0.314 lb), and atrazine (0.5 lb). All of the spray mixtures included liquid AMS at 2.5% v/v; all contained COC at 1% v/v except glyphosate, glufosinate, and fluroxypyr; and the dicamba and 2,4-D ester treatments contained NIS at 0.25% v/v. In addition, all treatments except glyphosate, contained 0.375 lb ae to control other weed species present. Visual weed control ratings were taken periodically throughout the growing season.

End of season ratings (approx. 8 WAT) revealed that glyphosate and the ALS-inhibitor (Group 2) herbicides provided only 47 – 67% marestail control. The contact and PPO (Group 14) herbicides had variable results. Bentazon provided only 50% control while the PPO herbicides ranged from 8 – 13% except for saflufenacil which provided 81% and glufosinate gave 92% control. The PGR (Group 4) herbicides were variable and ranged from 62 to 98% control. The 0.175 and 0.35 lb dicamba premix provided 96 and 98% control respectively. Dicamba alone and 2,4-D provided 86 and 75% control respectively. Control from the other two PGR herbicides was <73%. Mesotrione, topramezone, bromoxynil, and atrazine all were <53% control. However, the pyrasulfotole + bromoxynil premix was rated at 93% marestail control.

In summary, although the average marestail height was greater than what is suggested for good control it was realistic to when most are sprayed. We assume better control might have been observed by some (possibly all of the Group 4 herbicides, atrazine, cloransulam and chlorimuron) but not all of the treatments if applied to smaller weeds. Furthermore, saflufenacil may have provided better control if MSO instead of COC was used as recommended on the product label but none of the other Group 14 herbicide provided acceptable control of marestail. Some of the treatments (e.g., glufosinate) may have been less active if applied earlier in the year due to cooler temperatures. Higher spray volumes and different nozzles could potentially influence control as well. In general, herbicide combinations were not evaluated in this study but certain tankmixes may have provided some enhanced control. We plan to expand this study to assess some of these factors and other herbicides in the future.

PENDIMETHALIN CONTROLS JOINT-HEAD ARTHRAXON (*ARTHRAXON HISPIDUS*). R.

Chandran*, M.L. Flessner, Q. Johnson, B.M. Loyd, and S. Starcher, West Virginia University, Morgantown, WV

ABSTRACT

Arthraxon hispidus (Thunb.) Makino (joint-head Arthraxon, small carpetgrass, joint-head grass) is a non-native annual grass becoming increasingly prevalent in Appalachian pastures. This invasive weed, capable of displacing desirable forages, is unpalatable to livestock and is difficult to be managed. A field experiment was conducted in an established permanent cool-season pasture near Lost Creek (39.120425, -80.277615), West Virginia, in 2016, to evaluate the effect of pendimethalin applied at 4,484 g ai ha^{-1} (4 lb ai A^{-1}), preeemergence (PRE), and aminopyralid applied at 122 g ai ha^{-1} (0.109 lb ai A^{-1}), early postemergence (E-POST) to control joint-head Arthraxon. Pendimethalin applied on April 25^{th}consistently controlled joint-head Arthraxon >95% based on visual evaluations recorded periodically during the growing season, up to six months after treatment (MAT). Aminopyralid applied E-POST to the 2- to 3-leaf stage of joint-head Arthraxon provided 85% control until 1 MAT, but failed to control it later on; aminopyralid-treated plots were taken over by newly germinated joint-head Arthraxon subsequently. Pendimethalin applied PRE reduced shoot dry-weight of joint-head Arthraxon by 98% compared that in control plots (19.4 t ha^{-1}), at 6 MAT. No difference in shoot dry-weight of joint-head Arthraxon was recorded between plots that received aminopyralid (19.2 t ha^{-1}) and that in control plots.

FE-HEDTA SAFETY ON SEMI-DORMANT AND ACTIVELY GROWING ORNAMENTALS. J. Neal* and C.D. Harlow, North Carolina State University, Raleigh, NC

ABSTRACT

Experiments were conducted to evaluate the safety of late winter and summer applications of Fe-HEDTA over a diversity of woody and herbaceous ornamental plants. Plants were obtained from local nurseries about the first of March 2016, at which time most had not broken bud yet. *Hemerocallis* 'Stella de Oro', *Hosta* 'Minuteman', *Heuchera* 'Palace Purple', and *Sempervivum* were in 3- to 4-inch pots. *Sedum kamtschaticum, Sedum spurium* 'Red Carpet', and *Pachysandra* were in 2-inch cell-trays. *Juniperus* 'Andorra', *Gardenia* 'Radicans' and *Spirea* 'Gold Mound' were in 18-cell trays. *Liriope* 'Variegata' was in 3-L pots. For dormant applications, plants were arranged with 5 to 6 of each species per experimental unit (except 3 liriope per plot). Additional plants of those species were potted to 4-L pots and grown until early summer for a second experiment evaluating treatment to active growth. *Miscanthus sinensis* 'Gracillimus', *Pennisetum alopecuroides* 'Hameln', *Panicum virgatum, Muhlenbergia capillaris, Calamagrostis x acutiflora* 'Karl Foerster', *Schizachyrium scoparium* 'Prairie Blues', *Carex rosea,* and *Andropogon gerardii* were included for the summer treatments only. Treatments were arranged in a randomized complete block design with four replications. Fiesta (Fe-HEDTA) was applied three times at approximately 14 day intervals at 25, 50 or 100 oz/1000 ft^2. The dormant-season test was initiated on March 11th; and the summer experiment was initiated on June 9th. Crop injury was visually evaluated about 1, 3, 7, and 14 days after each application and 26 days after the last application. Injury was estimated using a 0 to 10 scale where 0 = no injury (indistinguishable from the non-treated) and 10 = dead plants or complete foliar necrosis. Intermediate values are estimates of % foliar necrosis and/or stunting. Pots with established bittercress, sowthistle and oxalis were included to assess weed control. Percent weed control was visually evaluated.

Dormant season applications caused no injury to juniper, gardenia, holly, or spirea. Pachysandra exhibited no foliar necrosis, but budbreak following treatment appeared to be delayed and warrants further study. Liriope foliage was discolored following 2 applications of 100 oz/1000 ft^2, but new growth emerging after treatment appeared to be normal. No significant injury was observed on dormant daylily, but when new growth was present, some necrosis was observed. Succulents (2 species of sedum and sempervivum) were severely injured. These results suggest that dormant season applications of Fiesta may have potential to control winter annual weeds in several woody nursery crops. In particular, conifers, holly, gardenia, spirea, and fully dormant herbaceous perennials exhibited no injury from dormant applications.

Summer applications to actively growing plants resulted in significant injury to all broadleaf ornamentals tested. However, no injury to juniper was observed. Injury to holly plants was less than 20% until the final evaluation, at which time plants treated with 100 oz/1000 ft^2 were smaller, suggesting the high doses of FeHEDTA were reducing growth. Daylily, hosta, heuchera, succulents, and pachysandra were severely injured. *Muhlenbergia* was not injured. Injury to *Calamagrostis* and *Schizachyrium* was less than 10% and temporary. Injury to other grasses was generally temporary. These data support continued evaluation of Fe-HEDTA for broadleaf weed control in conifers, ornamental grasses, and dormant woody ornamental crops.

CONTROL OF TURFGRASS WEEDS WITH TWO NEW HALAUXIFEN-METHYL CONTAINING FORMULATIONS (GF-3566 AND GF-2687) IN COOL AND WARM SEASON TURFGRASS. D.D. Loughner*, J.M. Breuninger, A.L. Alexander, and V.F. Peterson, Dow AgroSciences, Lawrenceville, NJ

ABSTRACT

GF-3566 is a systemic, postemergent herbicide composed of three proprietary active ingredients from Dow AgroSciences LLC for use on turfgrass. Two of the three active ingredients (halauxifen-methyl and 2,4-D choline) are new to the turf market and the third component is fluroxypyr. The three actives are synthetic auxin herbicides which act through a synthetic auxin mechanism (HRAC group O, WSSA group 4) mode of action.

GF-3566 provides quick activity and control of key problem weeds in cool season and bermudagrass turf. Upon US EPA registration, GF-3566 is expected to have a hazard signal word of "Warning" rather than the "Danger" signal word attributed to many of the 2,4-D amine containing products. The application rates will vary from 3.5 - 4.67 L/ha (3.5 - 4.0 pints/A) with use rates based on weeds and turfgrass species present. Positive attributes of GF-3566 include low odor and low volatility. GF-3566 is compatible with both low volume and traditional turfgrass application equipment, and mixes well in the tank with fertilizer and other products.

GF-2687 is also a systemic postemergent herbicide that controls both annual and perennial broadleaf weeds within southern turf stands. GF-2687 is a 1:1 ratio of halauxifen-methyl plus florasulam (HRAC group B, WSSA group 2) combining two distinct modes of action to help avoid and delay weed resistance. The application rate of GF-2687 is 50 g/ha (0.72 oz/A) and applications are rain fast after one hour. This low use rate provides effective weed control and is non-injurious across major warm and cool season turfgrass species. Turfgrass tolerance, even on herbicide sensitive St. Augustinegrass, has been demonstrated at temperatures above 32° C (90°F). Upon US EPA registration, GF-2687 is expected to have a Caution signal word with no buffer zone or temperature restrictions. Coupling these features with one rate, safety across numerous turfgrass species, and effective performance on targeted weeds, GF-2687 will deliver maximum application flexibility for turfgrass managers.

Tree studies have shown that GF-3566 and GF-2687 can be used under tree drip lines without concern for off-target or root uptake injury. Upon registration the expected use sites will include established turfgrass (commercial and residential), commercial sod farms, ornamental and sports turf, golf course fairways, aprons, roughs and tee boxes, campgrounds, parks, recreation areas, cemeteries, and unimproved turfgrass areas.

HALAUXIFEN-METHYL: A NEW, INNOVATIVE HERBICIDE FOR CONTROL OF BROADLEAF WEEDS IN TURFGRASS. J.M. Breuninger*, D.D. Loughner, A.L. Alexander, and V.F. Peterson, Dow AgroSciences, Indianapolis, IN

ABSTRACT

Halauxifen-methyl is a new herbicide for postemergent weed control in turfgrass, cereals and other crops and registrations for use on wheat and other cereal crops has been obtained in the U.S. and other countries around the world. Halauxifen-methyl is an innovative low-dose synthetic auxin (HRAC group O) herbicide and the first member of the new arylpicolinate class of chemistry, designed to provide unique attributes compared to other growth regulator herbicides. Halauxifen-methyl unique binding affinity in the cell nucleus differentiates it from previous synthetic auxin herbicides: Halauxifen-methyl demonstrates an affinity for the AFB5 auxin binding protein site of action in the cell nucleus of susceptible weeds.

Halauxifen-methyl provides consistent control of important broadleaf weeds in turf including common dandelion (*Taraxacum officinale*), narrow plantain (*Plantago lanceolata*), broadleaf plantain (*Plantago major*), common chickweed (*Stellaria media*), henbit (*Lamium amplexicaule*), and dollarweed (*Hydrocotyle sibthorpioides*). Trial work on both cool and warm season turf species including Kentucky bluegrass (*Poa pratensis*), perennial ryegrass (*Lolium perenne*) and tall fescue (*Festuca arundinacea*), bermudagrass (*Cynodon dactylon*), St. Augustinegrass (*Stenotaphrum secundatum*) and zoysiagrass (*Zoysia japonica*) has shown good turfgrass safety.

Halauxifen-methyl is effective at very low use rates of 10 g ae/ha and, due to its low vapor pressure, halauxifen-methyl does not cause off-target damage to desirable broadleaf plantings through volatilization. Tree studies on many different species have shown that the halauxifen-methyl can be used under the drip line without concern for off target injury. Halauxifen-methyl rapidly degrades in soils and plant tissues. Field and laboratory studies with Halauxifen-methyl were conducted at Purdue University, West Lafayette, IN and Woods End Research Laboratory, Mt. Vernon, ME to determine its fate in grass clippings and compost. It was determined that Halauxifen-methyl breaks down very quickly in turfgrass (DT_{50}=1.5 days) and has no significant or lasting herbicidal activity in compost. In November 2016, Dow AgroSciences submitted a request for registration to US EPA for Halauxifen-methyl containing formulations for use on turfgrass in both commercial and residential settings.

VEXIS: NEW TECHNOLOGY FOR TURF WEED CONTROL. J. Marvin* and A. Estes, PBI/Gordon, Kansas City, MO

ABSTRACT

PBI/Gordon would like to introduce a new active we are currently working on for US registration. Pyrimisulfan, a proprietary sulfonanilide, exhibits excellent safety on most turfgrass species as well as woody ornamentals. Pyrimisulfan is currently being formulated as a granule and liquid as an end use product.

Pyrimisulfan also demonstrates excellent herbicidal activity on plants in the *cyperaceae* family as well as some broadleaf weeds. Turfgrass safety has been observed at rates ≥360 g ai/ha, with both cool-season and warm-season turfgrass varieties exhibiting good tolerance.

Pyrimisulfan mode of action is acetolactate synthase (ALS) inhibition. Pyrimisulfan will have a maximum annual use rate of 90g ai/ha. Pyrimisulfan 35 – 90g ai/ha + Penoxsulam 35 – 90 g ai/ha has demonstrated excellent post emergence activity on a wide variety of broadleaf weeds. Control of yellow nutsedge (*Cyperus esculentus*), purple nutsedge (Cyperus *rotundus*) and false green kyllinga (*Kyllinga brevefolia*) has been observed at rates between 35 and 90 g ai/ha with single and sequential applications. In addition to post emergence activity, excellent preemergence activity has been observed on crabgrass, goosegrass and various broadleaf weeds. Length of efficacy appears to be rate dependent. Turfgrass safety on cool and warm-season is variable depending on rates, timing, and grass species.

Extensive research has been conducted throughout the United States. Planned uses include; home lawns, golf courses, sports turf, sod farms, industrial turf.

PREEMERGENCE CRABGRASS SPP. CONTROL IN TURFGRASS WITH EH1579 AND EH1580 CONTAINING VEXIS. G.M. Henry*, K.A. Tucker, J.T. Brosnan, G.K. Breeden, and A. Estes, University of Georgia, Athens, GA

ABSTRACT

Field experiments were conducted at the Athens Turfgrass Research and Education Center in Athens, GA and the East Tennessee Research and Education Center in Knoxville, TN during the spring and summer of 2016 to evaluate the preemergence control of crabgrass spp. (*Digitaria* spp.) with EH1579 and EH1580 containing Vexis. Both chemistries previously exhibited cool- and warm-season turfgrass tolerance. Research in GA was conducted on bermudagrass (5.1 cm) with a history of large crabgrass infestation [*Digitaria sanguinalis* (L.) Scop.], while research in TN was conducted on bermudagrass (1.6 cm) with a history of smooth crabgrass [*Digitaria ischaemum* (Schreb.) Schreb. Ex Muhl.]. The experimental design was a randomized complete block with four replications at each location. Herbicide treatments were initiated in GA on March 8, 2016 with a sequential application on May 4, 2016. Treatments consisted of EH1579 [90 g ai ha^{-1}, 180 g ai ha^{-1}, and 90 g ai ha^{-1} followed by (fb) 90 g ai ha^{-1}], EH1580 (90 g ai ha^{-1}, 180 g ai ha^{-1}, and 90 fb 90 g ai ha^{-1}), and indaziflam at 32.7 g ai ha^{-1} (industry standard). A non-treated check was added for comparison. Herbicide treatments were initiated in TN on March 7, 2016 with a sequential application on May 2, 2016. EH1579 and EH1580 treatments were similar in TN as previously described, but industry standards consisted of prodiamine at 840 g ai ha^{-1} and dithiopyr at 560 g ai ha^{-1}. Herbicides were applied with a CO_2 powered backpack sprayer calibrated to deliver 187 L ha^{-1}(GA) or 375 L ha^{-1} (TN). Visual ratings of % crabgrass cover was recorded 85 and 112 days after initial treatment (DAIT). Percent cover was compared to the non-treated check within each replication in order to determine % crabgrass control. EH1579 resulted in 52 to 56% large crabgrass control 85 DAIT when single applications were made at 90 and 180 g ai ha^{-1}. However, sequential applications of EH1579 at 90 g ai ha^{-1} resulted in 81% large crabgrass control 85 DAIT. EH1579 at 180 g ai ha^{-1} or 90 fb 90 g ai ha^{-1} resulted in 73 to 80% smooth crabgrass control in TN, while a single application of EH1579 at 90 g ai ha^{-1} only resulted in 43% control. Large crabgrass control was moderate in response to EH1580 85 DAIT. EH1580 at 180 g ai ha^{-1} or 90 fb 90 g ai ha^{-1} resulted in 58 to 62% large crabgrass control in GA, while a single application of EH1580 at 90 g ai ha^{-1} only resulted in 40% control. EH1580 at 180 g ai ha^{-1} or 90 fb 90 g ai ha^{-1} resulted in good smooth crabgrass control (89 to 93%), while EH1580 at 90 g ai ha^{-1} only resulted in 63% control 85 DAIT. Prodiamine at 840 g ai ha^{-1}, dithiopyr at 560 g ai ha^{-1}, and indaziflam at 32.7 g ai ha^{-1} resulted in 99%, 100%, and 63% crabgrass spp. control 85DAIT. Control of crabgrass spp. was $\leq$ 56% 112 DAIT in response to EH1579, regardless of rate, application frequency, or trial location. EH1580 at 180 g ai ha^{-1} or 90 fb 90 g ai ha^{-1} resulted in 68 to 69% large crabgrass control in GA, while a single application of EH1580 at 90 g ai ha^{-1} only resulted in 41% control 112 DAIT. Smooth crabgrass control was 84% 112 DAIT in TN in response to EH1580 at 180 g ai ha^{-1}, while control dropped to 63 and 43% in response to EH1580 at 90 fb 90 g ai ha^{-1} and 90 g ai ha^{-1}, respectively. Crabgrass spp. control was 96%, 91%, and 69% 112 DAIT in response to prodiamine at 840 g ai ha^{-1}, dithiopyr at 560 g ai ha^{-1}, and indaziflam at 32.7 g ai ha^{-1}, respectively.

JOINTHEADGRASS CONTROL IN COOL-SEASON TURF. J.M. Craft*, J.R. Brewer, and S. Askew, Virginia Tech, Blacksburg, VA

ABSTRACT

Jointhead grass *(Arthraxon hispidus (Thunb.) Makino)* is a low-growing sprawling, annual grass native to East and Southern Asia. Since its introduction in the 1970's, it has been reported in 28 states with the majority being located in the eastern United States. It is primarily a weed in pastures, ditches, roadsides, and turfgrass settings. Little to no previous research has been conducted examining jointhead grass control options in a cool-season turf setting. Therefore, research was conducted to evaluate commonly used postemergence herbicides for control of jointhead grass and safety on tall fescue (*Schedonorus arundinaceus* (Schreb.) Dumort., nom. cons). Two separate trials were initiated in Blacksburg, VA during the summer of 2016. The first trial was initiated June 10, 2016 on a tall-fescue roadside. The second trial was initiated July 14, 2016 on lawn height tall fescue. Treatments for both trials were arranged as a randomized complete block design with three replications. At 4 weeks after treatment (WAT), mesotrione at 0.28 kg ai ha^{-1} applied twice at a three-week interval, MSMA at 2.24 kg ai ha^{-1} applied twice at a two-week interval, fenoxaprop at 0.14 kg ai ha^{-1}, topramezone at 0.024 kg ai ha^{-1} applied twice at a three-week interval, and indaziflam at 0.067 kg ai ha^{-1} controlled jointhead grass 80% or more. Sulfentrazone at 0.28 kg ai ha^{-1} initially discolored jointhead grass but regrowth began within 3 WAT. While MSMA controlled jointhead grass, it injured tall fescue 75% and significantly reduced turf quality and turf cover. At 7 WAT, plots treated with MSMA had recovered and no injury was visible. Data from the two sites indicate that among herbicides registered for use in lawn turf, fenoxaprop, mesotrione, and topramezone can selectively control jointhead grass in tall fescue turf. Indaziflam and MSMA also control jointhead grass but are either injurious to tall fescue turf or not registered. The data further suggest that dithiopyr, fluazifop, quinclorac, and sulfentrazone may suppress but do not control jointhead grass in tall fescue turf.

HERBICIDE PROGRAMS FOR CREEPING BENTGRASS (*AGROSTIS STOLONIFERA*) CONTROL IN KENTUCKY BLUEGRASS (*POA PRATENSIS*). M.T. Elmore*, J.A. Murphy, B.S. Park, C. Mansue, and S. Alea, Rutgers University, New Brunswick, NJ

ABSTRACT

Experiments were initiated in October 2015 and April 2016 to evaluate various herbicide programs for selective creeping bentgrass (CBG; *Agrostis stolonifera*) control in Kentucky bluegrass (*Poa pratensis*). Treatments included: two and three sequential mesotrione applications at 280 and 175 g ha^{-1}, respectively, on three-week intervals or four applications at 140 g ha^{-1} on two-week intervals; tank-mixtures of three sequential mesotrione (175 g ha^{-1}) + triclopyr ester (280 and 560 g ha^{-1}) applications on three week intervals; four sequential applications of mesotrione (140 g ha^{-1}) + amicarbazone (50 g ha^{-1}) and mesotrione (140 g ha^{-1}) + amicarbazone (100 g ha^{-1}) applied at two-week intervals; two or three sequential applications of topramezone (37 g ha^{-1} or 32 g ha^{-1}, respectively) + triclopyr ester (560 g ha^{-1}) on three week intervals.

Both experiments were conducted on stand of mature 'Midnight II' Kentucky bluegrass naturally infested with CBG on a sandy loam soil at the Rutgers Plant Biology and Pathology Research and Extension Farm in Freehold, NJ. Treatments were arranged in a randomized complete block design with four replications and applied to 0.9 by 3.0 m plots using a CO2-powered single nozzle boom with 374 L ha^{-1} of water carrier and non-ionic surfactant (0.25% v/v) though an AI9504EVS nozzle. The first applications were made on October 15, 2015 and April 28, 2016 for the first and second experiment, respectively. Percent CBG reduction was determined by making an initial cover rating at the beginning of each experiment on a plot-by-plot basis and transforming subsequent CBG cover values to a percent reduction based on initial cover rating. Kentucky bluegrass injury was evaluated on a 1 (i.e., complete injury or death) to 9 (i.e., no injury) scale.

All treatments for the first experiment initiated in October 2015 reduced CBG similarly by > 70% at 5 months after initial treatment (MAIT) in March 2016. All treatments containing mesotrione reduced CBG similarly by 65 to 94% in June 2016. By September 2016, 11 MAIT, mesotrione tank-mixtures containing amicarbazone or triclopyr ester reduced CBG 74 to 87%, although reductions provided by treatments containing mesotrione alone (50 to 64%) were similar. Treatments containing topramezone reduced CBG < 50% at 11 MAIT. All treatments reduced CBG cover compared to the non-treated control; CBG cover increased by 15% in the non-treated from the beginning of the experiment to 11 MAIT. Treatments containing amicarbazone injured KBG at 3 and 7 weeks after initial treatment only.

For the second experiment initiated in April 2016, amicarbazone + mesotrione tank-mixtures reduced CBG by >80% at 2 MAIT; lesser CBG reductions were observed for all other treatments (< 50%) at 2 MAIT. By 5 MAIT, amicarbazone (50 g ha^{-1}) + mesotrione tank-mixtures reduced CBG by 89% although this treatment caused severe Kentucky bluegrass injury from 4 to 10 weeks after initial treatment. Other treatments containing mesotrione reduced CBG similarly by 16 to 50% at 5 MAIT. Topramezone did not reduce CBG cover compared to the non-treated control at 5 MAIT. Creeping bentgrass cover increased by 57% from the beginning of the experiment to 5 MAIT in the non-treated control in the second experiment.

HERBICIDE OPTIONS FOR ANNUAL GRASS WEED CONTROL DURING BERMUDAGRASS SPRIG ESTABLISHMENT. S. Askew*, J. Craft, and J.R. Brewer, Virginia Tech, Blacksburg, VA

ABSTRACT

Summer annual grasses, such as goosegrass and smooth crabgrass, are particularly troublesome during bermudagrass establishment from sprigs. Recent restrictions on MSMA use have left turfgrass managers with few options for weed management during bermudagrass establishment. A study was conducted in summer of 2016 to evaluate several herbicide programs for goosegrass, smooth crabgrass, and tufted lovegrass control and response of 'Latitude' bermudagrass during establishment from sprigs. Sprigs were row planted on July 12, 2016 and herbicide treatments were initiated 6 wk later. At 4 WAT, Drive XLR8 at 64 oz/A once or 32 oz/A twice at a 3-wk interval and Tenacity at 2 oz/A + Sencor at 4 oz/A controlled smooth crabgrass 81% or greater and equivalent to the hand-weeded check. Revolver at 17 oz/A alone or mixed with Drive at 32 oz/A, Tribute Total, Speedzone, Solitaire, and Pylex + Drive controlled smooth crabgrass 73% or less and less than the hand-weeded check. All foramsulfuron-containing treatments and Pylex at 0.5 oz/A + Drive at 32 oz/A controlled goosegrass 85% or more and equivalent to the hand weeded check 4 WAT. No treatment controlled tufted lovegrass greater than 60% or equivalent to the hand-weeded check at 4 WAT. Although the hand weeded check controlled all weeds 100% for the first 4 weeks, by 8 WAT some weed recovery or new emergence had reduced control to 85 and 87% for smooth crabgrass and goosegrass, respectively. At this timing, only Drive alone or Revolver at 27 oz/A + Drive at 32 oz/A controlled smooth crabgrass equivalent to the hand-weeded check. Only Revolver at 27 oz/A alone or with Drive, Tribute Total at 3.2 oz/A, and Pylex + Drive controlled goosegrass equivalent to the hand-weeded check. No herbicide controlled tufted lovegrass more than 43%. At 1 WAT, Revolver-containing treatments, Tribute Total, and Dismiss at 8 oz/A + Sencor at 4 oz/A were generally the least injurious (<27%) and treatments that contained Pylex (90%), Tenacity (47%), Drive (35 - 42%), or Speedzone(53 - 55%) were generally the most injurious. At 4 WAT, Drive applied twice and Pylex were the only treatments that injured bermudagrass more than 30%. At 5 WAT and beyond, no treatment injured bermudagrass more than 16%. Sheer strength at 4 WAT ranged from 57 to 66 n m^{-2} and was equivalent for all plots except Pylex + Drive, which was slightly lower at 54 n m^{-2}. These data suggest that Tribute Total or high rates of Revolver mixed with Drive cause the least bermudagrass injury while maintaining acceptable weed control of smooth crabgrass and goosegrass. Tenacity + Senor was initially injurious but bermudagrass quickly recovered and this treatment maintained a moderate level of weed control throughout the study and represents a more economical option. The Pylex rate used in this study was excellent for goosegrass control but too injurious to bermudagrass. Future work will seek to further refine treatment programs using the more successful herbicides in the current study.

THREE YEARS OF RESEARCH ON JAPANESE STILTGRASS CONTROL IN MANAGED TURF: WHAT HAVE WE LEARNED? J.R. Brewer*, S. Rana, and S. Askew, Virginia Tech, Blacksburg, VA

ABSTRACT

Previous research in native areas has shown that Japanese stiltgrass (*Microstegium vimineum*, JSG) is susceptible to nonselective herbicides like glyphosate and selective graminicides like sethoxydim and quizalofop. Few studies have evaluated JSG control in managed turf systems or when using selective herbicides registered for home lawns and other managed turf areas. At Virginia Tech, we have conducted multiple trials over three years to evaluate JSG control in cool-season turfgrass mown at lawn height. In 2014, no treatment injured turf greater than 20% at any rating. In 2015, four treatments injured turf between 30 and 48%, including mesotrione at 280 g ai ha^{-1} applied once (1X), mesotrione at 280 g ai ha^{-1} + triclopyr at 1120 g ai ha^{-1} 1X, mesotrione at 140 g ha^{-1} applied twice at a 3-wk interval (2X), and topramezone at 50 g ha^{-1} + triclopyr 1X. The difference in injury observed between the two years could be partially due to differences in moisture and temperature at each site. In 2014 at 6 WAIT, all rates of fenoxaprop (140, 70, and 35 g ai ha^{-1}) controlled JSG greater than 95% while topramezone at 25 g ai ha^{-1} 2X and topramezone + triclopyr 1X controlled JSG 65 to 68% and all other treatments controlled JSG less than 48%. In 2015, all fenoxaprop and topramezone-containing treatments controlled JSG greater than 95% while all other treatments controlled JSG less than 70%. JSG shoots were counted at the conclusion of both trials. In 2014, all rates of fenoxaprop contained less than 10 shoots per plot, and all other treatments had more than 240 shoots. In 2015, all topramezone and fenoxaprop-containing treatments contained less than 20 shoots per plot while all other treatments had greater than 250 shoots. In a separate study, only fluazifop caused unacceptable injury of 48% to Kentucky bluegrass. Fenoxaprop + fluroxypyr + dicamba (Last Call) at 421 g ai ha^{-1}, quinclorac + sulfentrazone + dicamba + 2,4-D (Q4) at 1.7 kg ai ha^{-1}, and fluazifop (Ornamec) at 53 and 105 g ai ha^{-1} controlled JSG greater than 95% by the end of the trial. Last Call and Q4 injured JSG more quickly than Ornamec. In 2016, the most effective treatments were fenoxaprop (70 g ai ha^{-1}) and topramezone (25 g ai ha^{-1}) regardless if applied in June or August. At both application timings, fenoxaprop controlled JSG greater than 90% by trial completion. Topramezone controlled JSG between 60 and 70% initially for both timings, but the later application allowed for less regrowth by first frost compared to the early timing (50 and 10% final control, respectively). In a separate 2016 study, sulfentrazone was applied at 5 rates between 0 and 421 g ai ha^{-1} but only rates at or above 281 g ha^{-1} transiently injured (between 70 and 80%) JSG 2 WAT. Q4 was also included in this study, but unlike the trial in 2015, Q4 controlled JSG less than 40%. By 8 WAT, no treatment controlled JSG greater than 5%.Differences in Q4 performance between years and transient JSG injury in the sulfentrazone rate study demonstrate that sulfentrazone-containing products may be inconsistent at controlling JSG. These studies indicate that herbicides commonly used for grass control in lawn turf including quinclorac, mesotrione, and sulfentrazone do not control JSG. Fenoxaprop selectively controls JSG at rates of 35 g ai ha^{-1} or higher and topramezone controls or suppresses JSG at rates of 25 g ai ha^{-1} or greater.

USING HERBICIDES AND PERENNIAL RYEGRASS TO RENOVATE TURF DOMINATED BY ANNUAL BLUEGRASS DURING 2015-16. B.S. Park*, M.T. Elmore, S. Alea, C. Mansue, and J.A. Murphy, Rutgers, The State University of New Jersey, New Brunswick, NJ

ABSTRACT

In golf and sports turfs, annual bluegrass (*Poa annua* L.) is among the most problematic weeds. Given its ability to rapidly germinate and aggressively establish, perennial ryegrass (*Lolium perenne* L.) can be most effectively overseeded into existing turfs compared to other cool-season turfgrasses. The objective of this study was to assess the effectiveness of herbicide programs designed to suppress annual bluegrass in turf overseeded with perennial ryegrass. Studies were initiated on a loam during September 2015 and 2016 on a mature stand of annual bluegrass in North Brunswick, NJ. Separate test areas were core cultivated with soil reincorporated on 3 and 7 September 2015 and 2016, respectively. Treatments consisted of 13 herbicide programs and an untreated check arranged in a randomized complete block design with four replications. Programs involved applications of glyphosate, amicarbazone, mesotrione, and ethofumesate applied during September through December 2015 and 2016 using a CO_2 backpack sprayer calibrated to deliver 375 L H_2O ha^{-1}. 'Manhattan 5 GLR' perennial ryegrass was slit-seeded at 527 kg ha^{-1} on 14 September 2015 and 537 kg ha^{-1} on 16 September 2016. Herbicide phytotoxicity (1 to 9 scale; 9=no phytotoxicity) was visually assessed during autumn 2015 and 2016. Perennial ryegrass populations were assessed during 2016 after autumn 2015 herbicide programs. Glyphosate (0.6 kg ai ha^{-1}; 4 September 2015) followed by (fb) ethofumesate (2.3 kg ha^{-1}; 6 October, 3 November, and 2 December 2015) resulted in the greatest perennial ryegrass populations observed in April (100%), May (99%), June (95%), and July 2016 (94%). During autumn 2015 and 2016, herbicide programs that included four or five sequential applications of mesotrione during 15 to 30 October 2015 and 21 October to 4 November 2016 resulted in severe phytotoxicity. Results from herbicide programs applied during autumn 2015 suggest that applications of ethofumesate after overseeding was the most effective herbicide strategy to establish perennial ryegrass while suppressing annual bluegrass.

A RAPID DIAGNOSTIC ASSAY TO DETECT HERBICIDE RESISTANCE IN ANNUAL BLUEGRASS. J.T. Brosnan*, J.J. Vargas, G.K. Breeden, and E.H. Reasor, University of Tennessee, Knoxville, TN

ABSTRACT

Reports of herbicide resistance in annual bluegrass (*Poa annua* L.) are greater than any other weed species commonly found in turf. Traditional means of testing annual bluegrass phenotypes for herbicide resistance are labor intensive, costly, and time consuming. Rapid diagnostic tests have been developed to confirm herbicide resistance in weeds of agronomic cropping systems that correlate well with traditional whole plant bioassays. Research was conducted at the University of Tennessee in 2016 to determine if agar-based rapid diagnostic tests could be used to confirm herbicide resistance in annual bluegrass phenotypes harvested from golf course turf. Separate experiments were conducted using annual bluegrass phenotypes resistant to acetolactate synthase (ALS) inhibiting herbicides and glyphosate via target site mutation; an herbicide susceptible control was included in each for comparison. Single tiller plants were washed free of soil and transplanted into autoclavable polycarbonate plant culture boxes filled with 65 mL of murashigee-skoog media amended with glyphosate (0, 6, 12, 25, 50, 100, 200, or 400 µM) or trifloxysulfuron (6.25, 12.5, 25, 50, 75, 100, or 150 µM). Treatments were arranged in a completely randomized design with 50 replications and repeated in time. Mortality in agar was assessed 7 to 12 days after treatment (depending on herbicide) and compared to responses observed after treating 98 individual plants of each phenotype with glyphosate (560 g ha^{-1}) or trifloxysulfuron (27.8 g ha^{-1}) in an enclosed spray chamber. Fisher's exact test ($\alpha = 0.05$) determined that mortality in agar with 100 µM glyphosate was not significantly different than treating whole plants via traditional spray application. Similarly, mortality in agar with 12.5 µM trifloxysulfuron was not significantly different than spraying whole plants with herbicide. While work is on-going to determine if this agar-based test can be used to assess resistance to photosystem II inhibiting herbicides and other modes of action, current findings indicate that this method can reliably provide an assessment of annual bluegrass susceptibility to glyphosate or trifloxysulfuron in 12 days or less.

DIRECTED LIQUID APPLICATIONS OF FLUMIOXAZIN EVALUATED FOR PHYTOTOXICITY AND PRE- AND POSTEMERGENCE CONTROL. H.M. Mathers*, Mathers Environmental Science Services, LLC, GAHANNA, OH

ABSTRACT

Three formulations of herbicides with the active ingredient flumioxazin were compared to an untreated control with three species of actively growing ornamentals in three gallon (C1200) containers. The three species were *Forsythia* "Lynwood Gold', *Syringa pekinensis* and *Pieris japonica* 'Red Mill'. All herbicides were applied around the base of the plants without contacting the foliage. Evaluation for phytotoxicity were conducted at 2, 4 and 8 weeks after application. Evaluations for weed control were conducted at 4, 8 and 12 weeks after application. Efficacy was evaluated pre- and post- weed emergence as the primary focus of this trial was weed control. Post- emergence evaluations were targeted on weeds no larger than 2" at time of application. Six treatments with four replications were conducted for phytotoxicity, pre- and post-emergence efficacy and post-emergence efficacy. The six treatments were SureGuard 51 WDG (8 oz/ac and 12 oz/ac), Panther 4SC (8 oz/ac and 12 oz/ac) and V-10233 76 WDG (with pyroxasulfone) (Nufarm Americas, Alsip, IL). All treatments were l*iquid applications and applied via a CO_2 backpack sprayer delivering 25 gal/ac (R&D Sprayers*, Opelousas, LA 70570) equipped with 8002 vs nozzles (TeeJet, East North Avenue, Carol Stream, IL 60116) operated at 45-50 psi with nozzle *spacing 18" apart.* The trail was initiated May 15, 2016. Efficacy was measured on a 0 to 10 scale, where 10 represents complete control and $\geq$7 represents commercially acceptable weed control and 0 no control. Evaluations for phytotoxicity used a rated score of 0 to 10 scale, where 0 represents no phytotoxicity, $\geq$3 represents commercially unacceptable injury and 10 represents plant death. At the last evaluation date growth indexes (GI) were also calculated as GI=Pi (Ht)(r^2), where Ht. was final height, r was half of the average of W1+W2 (two perpendicular measurements taken of plant diameter) and Pi was "p". Efficacy was perfect in pre- and post-emergence experiments for the duration of the study. There was unacceptable phytotoxicity with all treatments and species at various evaluations dates. However, by the end of the study most species had grown out of any injuries.

UNINTENDED INJURY TO ORNAMENTALS CAUSED BY CONSUMER-ORIENTED EXTENDED CONTROL HERBICIDES. A. Senesac*, Cornell Cooperative Extension, Riverhead, NY

ABSTRACT

Over the last few years, there has been a proliferation of ready-to-use (RTU) products that offer consumers postemergence weed control and additional extended pre-emergent control. It is the herbicides in these pre-mixes that provide the extended control that may be easily misused, sometimes with disastrous results.

In 2015-16, a study was conducted at the Long Island Horticultural Research and Extension facility to evaluate the effect of four commercially available extended control products on several established ornamental species. The study evaluated the products applied directed to the base of four tree and shrub species established for three years in the field. The treatments consisted of an application at the suggested labeled rate and at twice that rate with three replications. Care was taken to make the applications so that minimal foliage was contacted. The soil type is a Riverhead sandy loam with 1-2% OM. The application was made at the end of the summer (Sept. 18, 2015). The ornamental species were: red maple (Acer rubrum), Japanese plum yew (Cephalotaxus harringtonia 'Fastigiata'), dwarf fothergilla (Fothergilla gardenia) and inkberry, (Ilex glabra 'Densa'). The plots were irrigated within four days of treatment after which the plants were left alone until the treatments were evaluated the following spring (June 10, 2016). The four products that were applied were either ready to use or a concentrate that was diluted according to the label instructions were:

- Roundup Extended Control W&G Killer Plus Weed Preventer II (glyphosate 18%, diquat 0.73%, imazapyr 0.3%) (EPA Reg. No. 71995-40).
- Ortho Groundclear Complete Vegetation Killer Concentrate (glyphosate 5%, imazapyr 0.08%) (EPA Reg. No.239-2657).
- Bayer Advanced Durazone Concentrate W&G Killer (glyphosate 20%, diquat 0.9%, indaziflam 0.09%) (EPA Reg. No. 72155-100).
- Spectracide Weed & Grass Extended Control (diquat dibromide 2.3%, oxyfluorfen 1.92% fluazifop-p-butyl1.15%, dimethylamine salt of dicamba, 0.77%,) (EPA Reg. No. 9688-8845).

The results of visual estimation of injury suggest that only one of the the four products caused serious injury to the ornamentals the spring following the application. Although all the plants were negatively affected by Ortho Groundclear, the most dramatic injury was to the red maple. The leaves were greatly miniaturized and the growing points were dead in many cases. It should be emphasized the the label instructions for Groundclear very clearly state that no applications should be made within twice the distance from the drip line of any tree or shrub. These results dramatically illustrate the need for that precaution. There was some visible injury from other treatments, especially at the higher rate, although it did not reach a level of statistical significance. The ingredient in the Groundclear that is mostly responsible for the injury is imazapyr. This is a potent member of the imidazoline herbicide family. Once the injury is observed in the plant, it is unlikely that there will be significant recovery. Usually plant removal and replacement are necessary. These results indicate how easy it is to cause severe damage by not reading and following the product label instructions.

BIATHLON EFFICACY AND PHYTOTOXICITY IN ANNUAL BEDDING PLANT BEDS VERSUS SPECTICLE G, FREEHAND AND SNAPSHOT. H.M. Mathers* and E.J. Beaver, Mathers Environmental Science Services, LLC, GAHANNA, OH

ABSTRACT

The industry needs rotation products that can be used in mulched and un-mulched landscape beds that provide efficacy on a variety of weed species but low phytotoxicity with common annual bedding plants. Therefore, our objectives were to compare Biathlon G (Oxyfluorfen + prodiamine) (OHP, Inc., Mainland, PA 19451) at 1X, 1.5X and 2X (100, 150 and 200 lb/ ac, respectively) applied to mulched and un-mulched beds to FreeHand 1.75G (dimethenamid-p + pendimethalin) (BASF Corporation, Research Triangle Park, NC 27709) at 1X (150 lb/ ac), Snapshot 2.5 TG (isoxaben + trifluralin) (Dow AgroSciences, LLC, Indianapolis, IN 46268) at 1X (200 lb/ac) and Specticle G (indaziflam) (Bayer Crop Science Inc, Research Triangle Park, NC) at 1X (200 lb/ac) in un-mulched beds. All four products were labeled for landscape use but some species were not labelled. To observe any growth and quality impacts on the turf, all trial beds were located adjacent to turf. A total of eleven treatments were evaluated including, mulch only and a weedy check. Phytotoxicity was evaluated with seven bedding plant species and efficacy with "native" weeds. Because all landscape beds were newly established from turf in spring 2016, the native weed seed bank was plentiful. The most efficacious treatments at 14 weeks after treatment (WAT) evaluated by rated scores and fresh weights, were Biathlon 150 lb/ac and 200 lb/ac with mulch. Biathlon 100 lb/ac was still above commercially acceptable for weed control at 14 WAT with mulch but not without. At 14 WAT, only the FreeHand without mulch provided commercially acceptable control comparable to the Biathlon 150 and 200 lb/ac without mulch. Mulch additions to Biathlon treatments extended efficacy and increased weed control on average by 20%. Ratings and fresh weights positively correlated for efficacy but not for phytotoxicity. The lack of correlation for phytotoxicity can be attributed to the sensitivity of individual bedding plant species to various herbicides. Salvia, Coleus and Marigold had no phytotoxicity above commercially acceptable with any treatment throughout the trial. The impatiens were most susceptible to Specticle as the trial progressed. Biathlon had the most phytotoxicity on Snapdragon. The Petunia 'Celebrity Blue' overall was the most sensitive species evaluated with injury occurring above commercially acceptable with all four herbicides.

WEED SEED PLACEMENT EFFECTS ON PREEMERGENCE HERBICIDE EFFICACY. J. Derr*, Virginia Tech, Virginia Beach, VA

ABSTRACT

Experiments were conducted to determine if weed seeding time affects results of preemergence herbicide trials conducted in containers. Three active ingredients with different modes of action (flumioxazin, indaziflam, pendimethalin) were evaluated, each being applied using a sprayable and a granular formulation. Flumioxazin was applied at 0.25 lb ai/A, indaziflam was applied at 0.045 lb ai/a, and pendimethalin was applied at 3.0 lb ai/A. Weed seed was added to containers either prior to herbicide application and mixed into the top half inch of pine bark, or added to the pine bark surface after herbicide application. One theory was that contact herbicides such as flumioxazin should provide better weed control of weeds when seed was added prior to application, forcing weed shoots to penetrate through treated growing medium. Root inhibitors, such as indaziflam and pendimethalin, would be expected to provide greater weed control if weed seed was spread after application, forcing weeds roots to penetrate through treated growing medium. Trials were conducted during summer using long-stalked phyllanthus (*Phyllanthus tenellus* Roxb.), spotted spurge [*Chamaesyce maculata* (L.) Small], and southern crabgrass [Digitaria ciliaris (Retz.) Koel.], and in the fall using annual bluegrass (*Poa annua* L.), common chickweed (*Stellaria media* L.), and annual sowthistle (*Sonchus oleraceus* L.) to determine any difference between grasses and broadleaf weeds. In the spring and fall 2014 trials and in the spring 2015 trial, no interaction was seen between herbicide treatment and weed seeding time for any species. In the fall of 2015, no interaction between herbicide treatment and seeding time was seen for annual bluegrass and annual sowthistle. For common chickweed in 2015, indaziflam liquid and granular forms, and the granular formulations of flumioxazin and pendimethalin gave greater control when weed seed was added after application compared to before application, while sprayed flumioxazin and sprayed and granular forms of pendimethalin gave excellent control at both seeding timings. Overall, it does not appear that weed seeding time impacts contact herbicides differently than root inhibitors in container trials. For the nontreated checks, weed density tended to be similar for the two seeding timings. Weed shoot weight in nontreated plots was generally higher, though, when the weed seed was applied prior to application, perhaps due to less desiccation of plants as the seed was mixed into the growing medium as opposed to surface-applied seed for the after-herbicide seeding. Weed control was generally greater when weed seed was added after herbicide application, probably a reflection of the lower weed growth when seed was applied after application. The sprayable formulation generally provided greater weed control than the granular form for each active ingredient, probably due to greater coverage of the growing medium surface.

DORMANT APPLICATIONS IN MATURE LANDSCAPE BEDS SPECTICLE G VERSUS FREEHAND AND SNAPSHOT. H.M. Mathers* and J.A. Beaver, Mathers Environmental Science Services, LLC, GAHANNA, OH

ABSTRACT

The overall objective of this study was to evaluate the utility of Specticle G as an over the top dormant application versus two major industry competitor products, i.e. Snapshot 2.5 TG (isoxaben + trifluralin) (Dow AgroSciences, LLC, Indianapolis, IN 46268) and FreeHand 1.75G (dimethenamid-p + pendimethalin, BASF Corporation, Research Triangle Park, NC 27709) with 10 common landscape ornamental plants in eight evaluations from trial initiation to end, 01/16/2015 to 06/16/2016. There were six treatments applied to each species with three replications. The treatments included an untreated control, Specticle G at 200, 400 and 800 lb/ac, FreeHand at 150 lb/ac and Snapshot 2.5 TG at 200 lb/ac. Osmocote Pro 17-5-11 (3-4 month) fertilizer was applied at 3 lb/1000 ft^2 of landscape bed in the early spring. Efficacy was generally high in this experiment. All applications were conducted in mature landscape beds which limited competition due to early full crown development in spring. Also, applications were performed before any weeds had germinated and in cooler temperatures which seem to be advantageous for indaziflam (past trial experience). All treatment efficacies were above commercially acceptable, with the exception of the control, at 5 MAT for *Coreopsis verticillata* 'Zagreb', *Echinacea purpurea, Pennisetum alopecuroides* 'Moudry', *Festuca glauca* 'Elijah Blue', *Perovskia atriplicifolia* and *Salvia* × *sylvestris* 'Mainacht'. When efficacy was lost in *Yucca filamentosa,* and *Ligustrum* 'Vicaryi', it was with FreeHand, Snapshot and/or 200 lb/ac Specticle G. The Specticle G 200 lb/ac rate provided commercially acceptable control in *Yucca* at 4 MAT, but not at 5 MAT. The Snapshot was the least efficacious treatment in the *Yucca*. FreeHand with *Ligustrum* 'Vicaryi' did not provide commercially acceptable efficacy at 4 MAT and was not different than the efficacy provided by Specticle G 200 lb/ac or the control at 4 MAT. The FreeHand and 200 lb/ac rate of Specticle G were the worst two treatments in *Ligustrum*; however, Snapshot was statistically as efficacious as Specticle G 400 and 800 lb/ac at 5 MAT in *Ligustrum*. In summarizing efficacy and phytotoxicity results, FreeHand is listed for four species and Snapshot for three species as the most phytotoxic treatments in this study. Unfortunately, FreeHand and Snapshot are also listed for two species as the least efficacious treatments. In addition, although Specticle 400 (two species) and 800 lb/ac (one species) are listed as the most phytotoxic treatments, they are never listed as the least efficacious. Also, the Specticle G 200 lb/ac rate is listed for two species as the least efficacious treatment; however, it is never listed as the most phytotoxic. Generally, Specticle G 1X rate was superior to FreeHand and Snapshot in efficacy and reduced phytotoxicity and needs to be promoted in the landscape market as a dormant season application on mature beds. The extension of efficacy with the Specticle G in applying it dormant is very impressive. After five months, all the species had filled in well in the landscape and few weeds emerged as the season progressed through into fall.

UPDATE ON 2016 WEED SCIENCE RESEARCH IN THE IR-4 ORNAMENTAL HORTICULTURE PROGRAM. C. Palmer*, J. Baron, and E. Vea, IR-4 Project, Princeton, NJ

ABSTRACT

In 2016, the IR-4 Ornamental Horticulture Research Program sponsored research on two weed science projects: 1) crop safety of over-the-top in-season herbicide applications, and 2) crop safety of over-the-top herbicides applied to ornamental grasses. The goal of the over-the-top herbicide application evaluations was to screen herbicides [Biathlon (oxyfluorfen + prodiamine), Dimension 2EW (Dithiopyr), F6875 4SC (sulfentrazone + prodiamine), Freehand G (dimethenamid-p + pendimethalin), Gallery SC (isoxaben), Pendulum 2G (pendimethalin), Ronstar 2G (oxadiazon), SP1770, and Tower EC (dimethenamid-p)] for safety on woody and herbaceous perennials grown primarily in container nurseries with the emphasis on liquid herbicides. Applications were made at dormancy and approximately 6 weeks later for all products. Biathlon was applied to 23 crops; Dimension EW was applied to 12 crops; F6875 4SC was applied to 3 crops; Freehand was tested on 12 crops; Gallery was applied to 3 crops; Pendulum G was applied to 6, Ronstar was applied to 5 crops, SP1770 was applied to 5 crops, and Tower was applied on 22 crops. The goal of the ornamental grass herbicide product evaluations was to determine crop injury with over-the-top applications. Three products were evaluated: Dimension 2EW (Dithiopyr), Gallery SC (isoxaben), and Pendulum G (pendimethalin). These products were applied to 5, 14, and 7 species, respectively. The results from this research will aid in the development of product labels and will help growers and landscape care professionals make more informed product choices.

UTILIZING GLOBAL RESOURCES TO REGISTER NEW PEST CONTROL PRODUCTS ON SPECIALTY CROPS. D. Kunkel*, M. Braverman, and J. Baron, IR-4 Project, Princeton, NJ

ABSTRACT

IR-4 is recognized as the model publically funded program to help specialty crop growers obtain legal access to safe and effective pest management technology. IR-4 has developed numerous partnerships with other countries to share the burden of developing regulatory data to register products for minor uses and to create other "IR-4 like" programs. IR-4 and its partners realize that this cooperation is critical in finding solutions to the "Minor Use Problem" and to maintaining pest control options to growers. The products registered by IR-4 are often of the latest technology and important to IPM programs and to managing pesticide resistance, as well as managing new invasive pests. IR-4 has also supported capacity building for residue research projects in the ASEAN region, sub-Sahara Africa and in Latin America. It is anticipated that these partnerships will lead to collaborative research projects that can address the minor use needs of a broader community. These collaborations are also implementing specific data extrapolation proposals that allow residue data developed on certain representative crops to be used to support registration on a larger number of crops within a group (crop groups), which further assist specialty crop growers. Specialty crop producers around the world want access to the safest (newest) pest control tools that fit well into IPM programs that manage resistance and contribute to long term sustainability of both production and pest management.

ARE THERE ANY NEW HERBICIDES FOR SWEET CORN? D. Lingenfelter* and M. VanGessel,
Penn State University, University Park, PA

ABSTRACT

Field studies were conducted in 2016 at two locations, Rock Springs, Pennsylvania and Georgetown, Delaware, to examine various herbicide products in reduced-till sweet corn (*Zea mays succharata*, var. 'BC0805') that are newer to the market to determine their effectiveness on annual weed control. PRE and PRE fb POST programs were evaluated. PRE only treatments included: s-metolachlor + atrazine + mesotrione premix (2.48 lb ai/A); s-metolachlor + atrazine + mesotrione + bicyclopyrone premix (2.15 lb ai); saflufenacil + dimethenamid-P premix (0.65 lb) plus atrazine (1 lb); pyroxasulfone (0.133 lb) plus atrazine (1 lb); whereas, s-metolachlor + atrazine premix (1.38 lb); pendimethalin (1.43 lb); and pyroxasulfone (0.106 lb) were applied PRE followed by a POST application of one or a combination of the following herbicides: topramezone (0.0164 lb); topramezone + dimethenamid-P premix (0.67 lb); mesotrione + fluthiacet premix (0.094 lb); nicosulfuron (0.275 and 0.576 lb); tolpyralate (0.026 lb); mesotrione (0.094 lb); glufosinate (0.4 lb); and atrazine (0.5 lb). Necessary adjuvants were included in the POST spray mixtures. Visual weed control evaluations were taken periodically throughout the growing period. Sweet corn yield data and crop phytotoxicity ratings were also collected (data not included). Small-plot studies were arranged in a randomized complete block design with three replications.

In Pennsylvania, all treatments provided: >85% giant foxtail (Setaria faberi) control; >95% large crabgrass (Digitaria sanguinalis) control except for the nicosulfuron + topramezone mix (<50%); >90% common lambsquarters (Chenopodium album) control; >90% control of velvetleaf (Abutilon theophrasti) except for the nicosulfuron + topramezone tankmix (79%); >97% control of common ragweed (Ambrosia artemisiifolia); >98% redroot pigweed (Amaranthus retroflexus) control; and >85% control of common cocklebur (Xanthium strumarium) except for s-metolachlor + atrazine + mesotrione + bicyclopyrone premix, saflufenacil + dimethenamid-P premix plus atrazine, and pyroxasulfone plus atrazine (57-83%).

In Delaware, all treatments provided: >93% control of Palmer amaranth (Amaranthus palmeri); and >95% control of large crabgrass (Digitaria sanguinalis) except for topramezone + dimethenamid-P premix (70%) and nicosulfuron + topramezone tankmix (80%). Annual morningglory (Ipomoea spp.) was variable across treatments; most provided 88 to 99% control however, s-metolachlor + atrazine + mesotrione premix, pendimethalin fb glufosinate, and the nicosulfuron + topramezone tankmix ranged from 31 to 68% control. In general, crop injury was negligible for all treatments across both locations.

In summary, there are some new herbicide products that are a welcome addition for weed control in sweet corn. Many are simply premixes with existing active ingredients already labeled for use in this crop. Of more importance is getting the proper mix and rates of chemicals to not only provide effective weed control but also to account for potential carryover to rotation crops. For example, a lack of atrazine in the topramezone treatments was evident with reduced grass and morningglory control in some cases. In addition, late season morningglory control was impacted by not including full rates of atrazine in the PRE programs. Herbicides such as atrazine, mesotrione, topramezone, and pyroxasulfone potentially can leave residues causing injury to rotational crops. However, these can vary depending on use rates, soil types, rainfall, and other environmental conditions. A proper understanding of herbicide characteristics, herbicide resistance management techniques, and management of rotational crops can help alleviate crop injury yet obtain good weed control and improved crop yields.

PREEMERGENCE AND POSTEMERGENCE APPLICATIONS OF SULFENTRAZONE AND CARFENTRAZONE FOR WEED CONTROL IN WILD BLUEBERRY (*VACCINIUM ANGUSTIFOLIUM*) FIELDS. D.E. Yarborough* and J. D'Appollonio, University of Maine, Orono, ME

ABSTRACT

Zeus Prime XC (carfentrazone+sulfentrazone) is a product which has blueberry on the label but has not been tested in wild blueberries (*Vaccinium angustifolium* Ait.). Since no Group 14 products are registered for wild blueberry, Zeus can contribute to a weed resistance management program. Zeus was evaluated at different rates and timings on two pruned fields in Jonesboro and Wesley, ME. A Randomized Complete Split Block Design (four replications) was split with hexazinone (Velpar) applied pre-emergence 1 lb/a to half of each block. Treatments included: untreated check; Zeus 7.7 oz/a pre-emergence *(Zeus Low)*; Zeus 12.5 oz/a pre-emergence *(Zeus Mid)*; Zeus 15.2 oz/a pre-emergence *(Zeus High)*; Zeus 7.7 oz/a pre-emergence plus Zeus 7.5 oz/a post-emergence in November *(Zeus pre+fall)*; Zeus 7.7 oz/a pre-emergence plus rimsulfuron (Solida) 4 oz/a pre-emergence *(Zeus+Solida pre)*; Zeus 7.7 oz/ pre-emergence plus Solida 4 oz/a post-emergence *(Zeus pre+Solida post)*; and carfentrazone (Aim) 2 oz/a pre-emergence plus Solida 4 oz/a post-emergence *(Aim+Solida)*. Sites were evaluated in June and August. Treatments were compared to each other with Velpar treatments and no-Velpar treatments analyzed separately. Velpar vs no-Velpar for each main treatment and Zeus alone or with Velpar at four rates were compared on broadleaf and grass cover. Significant relationships of the latter were analyzed by regression. Phytotoxicity to wild blueberry was observed mainly as stunting and was observed with the Solida applications and with Aim. Aim+Solida controlled broadleaf weeds including problem weeds such as red sorrel (*Rumex acetosella* L.) and bladder campion (*Silene vulgaris* (Moench) Garcke), but with more injury to wild blueberry. Broadleaf weed control by Zeus released grasses, and the low Zeus rate did not control grasses well. In Wesley grass cover was primarily witchgrass (*Panicum capillare* L.), a later germinating species. Higher phytotoxicity expressed as stunting of the wild blueberry produced more grasses and the blueberry plants did not fill in as quickly or grow as tall as untreated. Zeus pre+Solida post Velpar vs no-Velpar treatments had comparable phytotoxicity and no-Velpar had more broadleaf weeds in June. Neither treatment had grasses in June but Velpar had more grasses in August. The early season broadleaf weeds delayed filling in of the wild blueberry canopy. For grass cover in Wesley for Zeus Mid and Zeus pre+Solida post, the difference does not appear due to August broadleaf weed cover, as Velpar vs no-Velpar was roughly equal. It is more likely due to Velpar controlling broadleaf weeds early in the season, which resulted in more favorable conditions for germinating grasses. A quadratic relationship found in Jonesboro for the Zeus rate on broadleaf weed cover suggests a diminishing return for the higher rates from injury to the wild blueberry plants or release of other weeds. The regression analyses for increasing rates of Zeus on weed cover are not strong and should not be considered definitive. The overall lack of significant relationships found are attributed to the lack of replication and field variability, so future trials will employ fewer treatments and increased replication.

EVALUATION OF HERBICIDE PROGRAMS FOR DRY BEAN, LIMA BEAN AND SNAP BEAN IN NEW YORK. D.E. Telenko*, J.R. Kikkert, and R. Bellinder, Cornell Univeristy, East Aurora, NY

ABSTRACT

Field trials in dry bean, lima bean and snap bean were conducted in 2016 at the H.C. Thompson Vegetable Research Farm, Freeville, NY to evaluate efficacy and crop tolerance to pre-emergence herbicide programs. These included: *s*-metolachlor + fomesafen, pyroxasulfone, acetochlor, prometryn, suflentrazone, carfentrazone+ sulfentrazone, cloransulam, pyroxasulfone + fluthiacet-methyl, and clomazone. The field sites were previously planted to winter wheat with a history of weed species that included common lambsquarters (*Chenopodium album* L), redroot pigweed (*Amaranthus retroflexus*), common purslane (*Portulaca oleracea* L.), Pennsylvania smartweed (*Polygonum pensylvanicum* L.), hairy galinsoga (*Galinsoga quadriradiata Cav.*), and large crabgrass (*Digitaria sanguinalis* (L.) Scop.).

Snap bean 'Slender Pack' and kidney bean 'Cal Early' were planted on May 24, and lima bean 'Kingston Baby' was planted on May 25. Pre-emergence treatments of *S*-metolachlor at 1.08 kg ai ha^{-1} + fomesafen at 0.28 kg ai ha^{-1}, pyroxasulfone at 0.07 kg ai $ha^{-1}$and 0.13 kg ai ha^{-1}, acetochlor at 0.45 kg ai ha^{-1} and 0.90 kg ai ha^{-1}, prometryn at 0.84 kg kg ai $ha^{-1}$and 1.12 kg kg ai ha^{-1}, sulfentrazone at 0.18 kg ai ha^{-1}, Spartan Charge (carfentrazone+ sulfentrazone) at 0.11 kg ai ha^{-1} and 0.13 kg ai ha^{-1}, cloransulam at 0.15 kg ai ha^{-1}, Anthem Max (pyroxasulfone + fluthiacet-methyl) at 0.15 kg ai ha^{-1}, and clomazone at 0.28 kg ai ha^{-1} were applied in plots on May 25 in all three trials. Plots were visually rated throughout the season for percent control on a scale of 0 to 100, with 0 being no control and 100 being complete weed control. Snap beans were harvested on August 5, dry beans on August 12, and lima beans on September 1.

The 2016 growing season was extremely dry. Supplemental irrigation was applied weekly in an attempt to minimize these conditions. No crop injury was noted in the snap beans, lima beans, and dry bean for all herbicide treatments. Herbicide efficacy was similar across all three bean trials with *s*-metolachlor + fomesafen, pyroxasulfone, prometryn, suflentrazone, and pyroxasulfone + fluthiacet-methyl all providing season long control of broadleaves and grasses. Acetochlor at 0.45 kg ai ha^{-1} did not control of lambsquarters, redroot pigweed, common purslane, smartweed, hairy galinsoga, large crabgrass, bindweed starting July 8 as compared to other treatments in lima bean. Loss of residual activity with acetochlor was also noted in both the green bean and dry bean trials for lambsquarters, redroot pigweed, and large crabgrass starting July 8. Spartan Charge at both rates (0.11 and 0.13 kg ai ha^{-1}), cloransulam and clomazone had less residual activity for large crabgrass starting mid-season (July 22). No differences in yield were detected in both snap bean and kidney bean for all herbicide treatments. Yield of lima bean was greatest with Anthem Maxx, and was not significantly different from clomazone, cloransulam, and pyroxasulfone at 0.13 kg ai ha^{-1}. Lima bean yield was lowest under treatments where there was a loss of residual weed control in acetochlor 0.45 kg ai $ha^{-1}$and sulfentrazon 0.18 kg ai ha^{-1} treatments, but these were not significantly different from the handweeded check.

IMPROVING WEED CONTROL IN CRANBERRY WITH NOVEL USES OF REGISTERED HERBICIDES. K.M. Ghantous* and H. Sandler, UMass Cranberry Station, East Wareham, MA

ABSTRACT

Cranberry growers have a limited suite of herbicides registered for use, with only three gained within the past 20 years. To meet weed management challenges, we focused on novel uses of registered compounds. Massachusetts cranberry growers need options to control perennial grasses, especially broomsedge bluestem (*Andropogon virginicus;* BS) and little bluestem (*Schizachyrium scoparium;* LBS).

Napropamide is a preemergence herbicide that has been used in cranberry for over 35 years. Greenhouse trials have demonstrated that napropamide controls germination of BS and LBS seeds, however control is not seen when used in the field. It is currently labeled for application before spring growth begins (April or early May) or in the fall after harvest and the safety of growing season applications has not been evaluated. We hypothesized that herbicide activity from early spring applications do not overlap with the germination periods for BS and LBS, which are warm-season grasses. We evaluated the crop safety of napropamide 42.1L ha^{-1}; 10.01 kg a.i. ha^{-1}) applied in 3,742 L ha^{-1} of water to simulate application by chemigation (CH) at various stages of cranberry phenology (budbreak, roughneck, hook stage, bloom, fruit set). Each plot (1 m^2) received a single treatment and treatments were replicated four times. Plots were visually monitored throughout the season, and no injury was noted for any treatment. Cranberry fruit was collected from a 930 cm^2 quadrat from each plot at the conclusion of the study and evaluated for number and weight of sellable fruit. There were no differences between any treatment and the untreated control. Fruit samples will be analyzed for residue.

Clethodim is effective for postemergence control of BS and LBS. Clethodim products cannot be chemigated in cranberry, which is the primary method of pesticide application for Massachusetts cranberry growers. A study was conducted that compared broadcast applications (applied in 281 L ha^{-1} of water; BC) to CH applications of a clethodim product (1.17 L ha^{-1}, 12.8% a.i. with NIS at 0.25% v/v) to evaluate cranberry crop safety. Each plot (1 m^2) received a one of eleven treatments, and treatments were replicated four times: 1 application by BC or CH at roughneck, after bloom, and 14 days after bloom; 2 applications by BC or CH at roughneck + after bloom and after bloom + 14 days later, or untreated control. Plots were visually monitored throughout the season, and no injury was noted with the exception of some floral deformities in roughneck treatments. Cranberry fruit was collected from a 930 cm^2 quadrat from each plot at the conclusion of the study and evaluated for number of sellable fruit and weight of sellable fruit. There were no differences between any treatment and the untreated control. The novel use patterns of applying napropamide later in the season and chemigation of clethodim demonstrated crop safety for cranberry vines and fruit.

COLONIZATION BY NEWLY PLANTED CRANBERRY VINES IS INFLUENCED BY WEED MANAGEMENT, HORTICULTURAL AND CULTIVAR CHOICES. H. Sandler* and K.M. Ghantous, UMass Cranberry Station, East Wareham, MA

ABSTRACT

Management choices during the establishment of cranberry vines, including the use of preemergence and postemergence herbicides, may affect the time to first cropping and the long-term productivity of the planting. A single application of granular napropamide (Nap) at the maximum label rate of 3.72 kg ai ha^{-1} was the traditional recommended preemergence herbicide for new plantings due to its crop safety properties and efficacy when applied to weed-free surfaces. A new formulation of napropamide (dry flowable) and mesotrione (Mes) were labeled for use in cranberry in 2009 and 2008, respectively. The objective of this study was to evaluate the weed control provided by Nap (dry flowable) and Mes (emulsion) applied annually, as single, multiple or sequential combinations, over a 2-yr period on newly planted vines or cranberry plantings in their second year of growth.

Six locations were studied over a 4-yr period (2009-2012); each individual site was treated for a 2-yr period. Three sites were treated in the year of planting plus the subsequent year (called "new plantings"; NP) and three sites were treated in their second year of growth plus the subsequent year (called "second-year plantings"; SY). Ten treatments, delivered in 3,735 L ha^{-1} water, were administered each year: Nap at 3.36 kg ai ha^{-1} applied once, twice, thrice or once followed by (fb) one application of Mes at 210 g ai ha^{-1}; Nap at 5.04 kg ai ha^{-1}applied once, twice or once fb one application of Mes at 210 g ai ha^{-1}; Mes at 210 g ai ha^{-1} applied once or twice; and untreated. Weed and cranberry biomass were collected and plant community surveys were conducted annually for all sites.

Correlation analysis indicated that monocot biomass production was the primary positive predictor for total weed biomass and negatively related to cranberry biomass. At NP sites, multiple herbicide applications reduced monocot biomass more than a single application, regardless of herbicide choice. At the SY sites, Mes-only treatments reduced total weed biomass more than Nap-only treatments. No herbicide injury on the cranberry plants or impact of treatment on cranberry biomass production was observed but overall vine colonization was poor for several sites indicating that other factors, including cultivar choice, planting density, the use of rooted plugs, fertilizer, and water management, may play a larger role in rapid cranberry vine colonization than the suppression of initial weed biomass.

Since grasses were often the predominant early colonizers in young cranberry plantings and given the complementary range of efficacy for Mes and Nap, a combination PRE-POST herbicide program of a low rate of Nap fb Mes may be the best cost-benefit program in many instances. If cost is a limiting factor, applying a single application of Mes to a new planting should be included as a component of cranberry bed establishment since this practice consistently resulted in significantly less initial weed biomass compared to areas left untreated.

POTATO (*SOLANUM TUBEROSUM*) TOLERANCE AND WEED CONTROL BY FLURIDONE APPLIED PREEMERGENCE AFTER DRAG-OFF. C.W. Cahoon*, M.L. Flessner, and T. Hines, Virginia Tech, Painter, VA

ABSTRACT

Fluridone, a Group 12 herbicide, inhibits phytoene desaturase, an enzyme necessary in the biosynthesis of carotenoids. With norflurazon no longer sold for cotton, no Group 12 herbicides currently are used on agronomic crops in the U.S. Development of glyphosate- and ALS-resistant palmer amaranth renewed interest in using fluridone for residual control of this weed in cotton. Sweetpotato tolerance to fluridone is also reported to be good. Currently, potato producers rely primarily on metribuzin and s-metolachlor for preemergence weed control. However, there is growing concern over weeds developing resistant to triazine herbicides (metribuzin). To be proactive in avoiding herbicide resistant weeds, potato producers need an expanded portfolio of herbicides. The objectives of this research were to determine potato tolerance to fluridone and evaluate weed control by residual combinations including fluridone for use in potato. Two experiments (potato tolerance and weed control) were conducted at the Eastern Shore Agriculture Research and Extension Center near Painter, VA during 2016. Soil at the research site was sandy loam with 1% organic matter and pH 5.8. Common ragweed was the most prominent weed in the weed control study. Potato in both studies were planted on March 18, 2016. At 19 days after planting, hilling followed by drag-off operations were preformed just prior to applying residual herbicides preemergence. All herbicides were applied using a CO_2-pressurized backpack sprayer calibrated to deliver 20 GPA at 24 PSI. For the potato tolerance study, fluridone was applied after drag-off at 0, 0.1, 0.15, 0.2, 0.4, and 0.8 lb ai/acre. Fourteen days after preemergence (DAP), fluridone at 0.1, 0.15, 0.2, 0.4, and 0.8 lb ai/A injured potato 7, 19, 23, 42, and 66%, respectively. Little potato injury was observed by the 0.1, 0.15, and 0.2 lb ai/A rates of fluridone at 56 DAP. However, 0.4 and 0.8 lb ai/A of fluridone injured potato 37 and 62% 56 DAP. Potato yield in plots treated with 0.1, 0.15, 0.2, and 0.4 lb ai/A fluridone yielded similarly to non-treated plots. The 0.8 lb ai/A rate of fluridone reduced potato yield approximately 50 cwt/A.

For the weed control study, fluridone, fomesafen, metribuzin, and *s*-metolachlor were applied alone at 0.15, 0.25, 0.31, and 1.3 lb ai/A, respectively. Combinations of these herbicides also evaluated included fluridone + fomesafen, fluridone + *s*-metolachlor, fluridone + metribuzin, fomesafen + *s*-metolachlor, metribuzin + *s*-metolachlor, fluridone + fomesafen + *s*-metolachlor, fluridone + fomesafen + metribuzin, and fomesafen + metribuzin + *s*-metolachlor. At 42 DAP, fluridone alone, fomesafen alone, metribuzin alone, and *s*-metolachlor alone controlled common ragweed 72, 97, 75, and 23%, respectively. All combinations except fluridone + *s*-metolachlor (85%) controlled common ragweed 90% or better 42 DAP. Potato yielded 298 cwt/A or greater in all treatments except combinations that included both fluridone and metribuzin. Plots treated with fluridone + metribuzin and fluridone + fomesafen + metribuzin yielded 271 and 266 cwt/A, respectively.

Overall, potato tolerance to fluridone is good. However, additional studies conducted under varying environmental conditions are needed to determine if potato tolerance is acceptable. Fluridone provided similar weed control to commonly used potato herbicides. If potato tolerance to fluridone proves acceptable, potato producers will have another mode of action at their disposal; thus reducing the risk of metribuzin-resistance.

INVASIVE VASCULAR PLANT SPECIES AT CALEDON STATE PARK, VIRGINIA. R. Stalter*, St. John's University, Jamaica, NY

ABSTRACT

The objective of this study was to determine invasive taxa at Caledon State Park, Virginia, a 1044 ha park, 38 21' 09"N, 77 07' 58" W bordering the Potomac River. The study was initiated in May 2015 and will be terminated in October, 2017. Monthly Trips to the park were taken from March to October to collect vascular plant species. A total of 628 taxa including 118 non- native taxa have been identified. All spore plants (28 taxa) and gymnosperms (4 taxa) are native to the region. Thirty one monocots, 16.8 % of the 185 monocot taxa and 87 dicots, 20.2 % of the 411 dicot taxa are not native to the region. One hundred appropriately sized quadrats have been established at five woodland sites (20 quadrats/site) at Eagle Wildlife Area, Fern Hollow Trail (2 sites) Symplocarpus Ravine and Potomac River bluff to determine frequency of occurrence of vascular plant species. The most frequently encountered non- native woodland taxa present at one or more of the five sampled sites are *Microstegium vimineum* (100% frequency) *Rubus phoenicolasius* (75% frequency) and *Lonicera japonica* (25% frequency). These three woodland invasives, especially *M. vimineum* have the greatest impact on native taxa. Wetland borders of Jones Pond and Potomac River inlets are heavily colonized by non-native *Phragmitesaustralis*. This aggressive invasive may become more wide spread at Caledon's wetlands in the future

AMERICAN BURNWEED, BLACKBERRY, AND GROUND IVY CONTROL WITH AMINOCYCLOPYRACHLOR. J.R. Brewer*, J. Craft, and S. Askew, Virginia Tech, Blacksburg, VA

ABSTRACT

Method (aminocyclopyrachlor) is a potent herbicide utilized for pasture and industrial vegetative management (IVM) areas. Although much is known about weeds controlled by aminocyclopyrachlor in these systems, more weeds could be added to the product label given sufficient research testing. In the summer of 2016 at Virginia Tech, four different studies were conducted to evaluate aminocyclopyrachlor on ground ivy (*Glechoma hederacea*), American burnweed (*Erechtites hieraciifolius*), highbush blackberry (*Rubus pensilvanicus*), and black raspberry (*Rubus occidentalis*). The ground ivy site was initiated on April 18, 2016, on a perennial ryegrass nonmow area. Two studies on American burnweed and highbush blackberry were initiated on August 17, 2016 in an out-of-play area of the Virginia Tech Golf Course (VTGC) that receives one annual mowing. The fourth study included both highbush blackberry and black raspberry, which was also placed at the VTGC, and it was initiated on October 14, 2016. The treatments from the first three trials included: aminocyclopyrachlor at 70, 140, 211, and 281 g ai ha^{-1}, Milestone (aminopyralid) at 123 g ai ha^{-1}, and Crossbow (2,4-D + triclopyr) at 2.5 kg ai ha^{-1}. The fourth trial included: aminocyclopyrachlor at 0.5% v/v, aminopyralid at 0.5% v/v, 2,4-D + triclopyr at 1.5% v/v, and Escort (metsulfuron) at 0.18 g ai L^{-1}, and all treatments were mixed with 1% v/v of methylated seed oil.

Triclopyr + 2,4-D injured perennial ryegrass approximately 18%. All treatments controlled ground ivy and there was no difference observed by the rates of aminocyclopyrachlor. At 4 months after treatment (MAT), triclopyr + 2,4-D and aminopyralid treated plots were exhibiting signs of ground ivy recovery. All treatments controlled American burnweed greater than 80% by 1 MAT, except for the lowest rate of aminocyclopyrachlor. By 2 MAT, the same treatments had controlled American burnweed greater than 95%. The lowest rate of aminocyclopyrachlor controlled American burnweed less than 75% by 10 weeks after treatment (WAT). The highest rate of aminocyclopyrachlor also suppressed quackgrass (*Elytrigia repens*) greater than 50% 2 MAT. During midsummer, aminocyclopyrachlor controlled highbush blackberry less than was observed for other weeds. By 1 MAT, triclopyr + 2,4-D had controlled highbush blackberry 99%, while all other treatments controlled highbush blackberry less than 60%. By 10 WAT, the three highest rates of aminocyclopyrachlor controlled highbush blackberry approximately 60% compared to complete control by triclopyr + 2,4-D. When applied in fall, aminocyclopyrachlor, aminopyralid, and triclopyr + 2,4-D controlled both black raspberry and highbush blackberry greater than 80% 4 WAT, while metsulfuron controlled highbush blackberry 65% and black raspberry 15%. The difference observed between the two blackberry studies is possibly a result of two different application methods. The first study used a broadcast application method, while the second study used a spray-to-wet method. The final rating for the fall-initiated trial will be assessed in spring 2017.

COMPARING A WET-BLADE PRUNING DEVICE TO FOLIAR SPRAY FOR CHEMICAL MOBILITY IN WOODY STEMS. S. Askew*, T. Burch, S. Rana, and J.R. Brewer, Virginia Tech, Blacksburg, VA

ABSTRACT

A handheld pruning device that utilizes a wet-blade chemical application technology was invented by Tom Burch. The device combines pruning and chemical treatment operations for efficient management of unwanted weeds or systemic treatment of woody crop diseases. An individual plant treatment (IPT) study was conducted in Blacksburg, VA to evaluate systemic movement of chemical dye into tartarian honeysuckle (*Lonicera tartarica*) stem tissue. The study was conducted as a completely randomized design with four replications. Mature honeysuckle bushes with trunk diameters of at least 2 cm were randomly selected from a forest understory and marked for foliar spray or pruning treatment. Plants were treated with a nonionic polymeric colorant (Blazon®) by spraying all foliage on a selected branch or by pruning a selected branch with the wet-blade pruning device. In each case, the selected branch received 100 µl of colorant and no other branches on the plant were treated. Nondiluted colorant was ejected onto the cutting blade of the pruner as the branch was cut or mixed with 6 ml water and sprayed onto all foliage of one selected branch with a spray atomizer. Stem sections were collected 24 hr later at five distances that ranged from 0 to 40 cm below the treated stem. Bark was removed from each stem section and sections were imaged with a digital camera to record the top and bottom cross sections and four sides. The resulting six images were subjected to digital image analysis to determine the percentage of blue pixels detected. Increased phloem movement of the blue colorant was evidenced by blue streaks or stripes leading down the stem and in the outer rings of stem cross sections. The resulting percentage of blue pixels detected was regressed against distance from the treated stem. When treated with the wet blade pruner, 14% of imaged tissue was blue when stem sections were taken 0 to 2 cm below the treated stem and the percentage decreased in a curvilinear fashion to 2% blue pixels at 40 cm below the treated stem. When foliage was sprayed, only 1% blue tissue was detected at 0 - 2 cm below the treated stem and no blue pixels were detected beyond 20 cm below the treated stem. The data from this study suggest that chemical wiping of cut stems may be a valid method to treat woody vegetation when phloem movement is desired.

EFFECTS OF PHOSPHATE ON GLYPHOSATE UPTAKE, TRANSLOCATION AND TRANSPORT IN EUCALYPT: IMPLICATIONS FOR NUTRIENT MANAGEMENT. F. Campos Mastrotti Pereira, P. da Costa Aguiar Alves, and W.A. Peer*, University of Maryland, College Park, MD

ABSTRACT

Soluble phosphate availability is the major limiting factor for plant growth, development and yield. Plants evolved high and low affinity phosphate transporters as an adaptive mechanism to optimize phosphate acquisition. Phosphonomethyl glycine (glyphosate) is an herbicide widely used throughout the world, and previous studies have suggested that glyphosate can be transported across the plasma membrane via phosphate transporters. To investigate this hypothesis, experiments were conducted in *Eucalyptus grandis*. *Eucalyptus grandis* phosphate transporters showed differential gene expression in leaves and roots in response to phosphate deficiency. In general, the expression of high affinity phosphate transporters increased in phosphate starvation conditions, particularly in roots. ^{14}C-glyphosate absorption and translocation of ^{14}C-glyphosate was applied to leaves or roots in whole seedlings, and ^{14}C-glyphosate was absorbed at a higher rate and translocated faster in phosphate starved plants compared to controls grown in phosphate replete conditions. In leaf mesophyll protoplast assays, ^{14}C-glyphosate was rapidly taken up, and NaH_2PO_4 competed with ^{14}C-glyphosate uptake, indicating the glyphosate was taken up via the phosphate transporters. These results have implications for best management practices for weed control via glyphosate application under phosphate application regimes.

DOES WEED CONTROL NATTER IN REFORESTATION WITH TREE SEEDLINGS? A. Gover*, Penn State University, University Park, PA

ABSTRACT

Previous research conducted by this project investigating post-plant weed control effects on tree seedling establishment in unfertilized wildland settings provided results that were barely or not significant. Intensity of pre-plant weed suppression was variable in previous studies, as was planting stock source, and site conditions. In a effort to clarify vegetation management impacts, an experiment was established at two sites evaluating the interaction of pre-plant and post-plant weed suppression. A factorial design with two levels of pre-plant (with or without) and three levels of post-plant (0, 2, 4 years) weed suppression in plantings of tulip poplar (*Liriodendron tulipifera* L.) seedlings was established in riparian settings in Elverson and West Chester, PA. The Elverson site was an Edgemont channery loam (Fine-loamy, mixed, active, mesic Typic Hapludults; NRCS capability class 2e), and the West Chester site was a Codorus silt loam (Fine-loamy, mixed, active, mesic Fluvaquentic Dystrudepts; NRCS capability class 2w). The productivity of the two sites would be inherently similar, but the Elverson site appeared to be more eroded with less topsoil, while the West Chester site was a loamy, alluvial soil. The pre-plant treatment of glyphosate plus sulfometuron plus pendimethalin (2.2 plus 0.023 plus 4.4 kg/ha, respectively) was applied to both sites October 29, 2015, treating 1.8 by 1.8 m squares, arranged in a completely randomized design with eight replications. Containerized seedlings from a single nursery averaging 47 cm in height were planted April 26 and May 6 at Elverson and West Chester, respectively. The post-plant herbicide was applied immediately after planting, using the same mix as the pre-plant treatment. Height of tallest terminal bud and stem diameter was measured on all stems at each site November 10, 2016. Data for 2- and 4-yr post-plant treatments was combined for the analysis. There was no interaction between pre- and post-plant treatment for height or diameter for either site, and post-plant treatment was not significant for any dependent variable. At West Chester, pre-plant treatment was a significant effect for stem height (142 cm with, 105 cm without) and diameter (12.9 mm with, 11.3 mm without), while at Elverson there was no significant difference for stem height (73 cm with, 65 cm without) or diameter (10.3 mm with, 10.0 mm without). When tree growth at Elverson was characterized as percent height growth from planting, pre-plant weed control was a significant effect (101 percent with, 50 percent without). When compared with findings in the literature on seedling tree growth in intensively cultured orchard settings, these data suggest that weed suppression effects are correlated to growth conditions, with suppression of competition providing more effect on more fertile sites. One question this data raises is whether we should make a practice of fertilizing riparian tree plantings to improve growth and optimize response to weed suppression treatments.

DEMOGRAPHY OF POISON IVY IN FOREST INTERIOR AND EDGE HABITATS. C. Dickinson*
and J.N. Barney, Virginia Tech, Laneview, VA

ABSTRACT

Poison ivy, *Toxicodendron radicans* (L.) Kuntze, is a native noxious weed with invasive tendencies, which causes contact dermatitis to millions ever year. To better understand the ecology of *T. radicans,* it is necessary to understand the germination, establishment, and habitat preferences of this vine. Locally collected seeds collected were planted in the understory or forest edge at five sites. Seeds were either carefully removed from the panicles and then planted ("Raw") or mechanically and chemically scarified to mimic passage through a bird gut, which is the putative primary fruit disperser. Sites were chosen for different plant communities and the lack of significant resident *T. radicans* populations. In early May 2016, we planted 25 seeds of either treated or raw seeds were planted in 10 $1m^2$ plots in each habitat, for a total of 2500 seeds planted. The emergence and survival of seedlings was tracked weekly from May-August and then monthly from September-October of 2016. Percent light transmittance, soil moisture, and percent cover of plant functional groups and bare ground data was collected per plot. Soil samples were collected for each habitat and analyzed by the Virginia Tech Soil Testing Lab. This data will be collected in the same fashion in 2017 and 2018 with one modification: weekly surveys of plant emergence will begin when resident poison ivy starts the growing season in Spring. At the culmination of the study, any remaining plants will be collected for above-ground biomass data. The initial germination results indicate that raw seeds do require cold stratification to break dormancy, contradicting the results of previous work. Further, survival surveys suggest that herbivory pressure of seedlings may be quite high. These results have laid the foundation for an herbivore exclusion study to characterize the herbivory of young *T. radicans* seedlings, and reopened the investigation of the mechanisms behind the dormancy of *T. radicans* seeds.

NEWSS YEAR-END FINANCIAL STATEMENT 2016

AUDIT REPORT – Erin Hitchner

Financial documents and bank accounts were audited by,
Kurt Vollmer and Quintin Johnson on December 7, 2016.

BEGINNING STATEMENT: As of October 31, 2015, the NEWSS checking account showed a balance of **$22,015.88**, while the savings account balance was **$31,838.21**. The 60 month CD balance was **$26,304.96**.

ENDING STATEMENT: As of October 31, 2016, the NEWSS checking account showed a balance of **$53,502.86**, while the savings account balance was **$31,844.57**. The 60 month CD matured on March 31, 2016 and a total of **$26,524.45** was deposited in the NEWSS checking account.

Net gain in assets for the fiscal year: $108,961.82 - $102,655.70 = $6,306.12

FISCAL YEAR ACCOUNTING

New income as shown on bank statements for November 2015 through October 2016 including earned interest was **$102,033.59.**

New deductions from checking for November 2015 through October 2016 totaled $**95,727.47.**

Net gain *in assets:* ***$102,033.59 - $95,727.47 = $6,306.12***

Non Annual Meeting Revenue/Expenses:

Non annual meeting revenue, which was comprised of interest earned as well as weed contest support from sustaining members and other contributors ($7,025.00) totaled $8,068.59

Non annual meeting expenses (Dishonesty bond, website maintenance, Apex abstract fee (2015 & 2016), weed contest, administrative costs, WSSA Rep Travel, Dir. Science Policy (2015 & 2016) and CAST dues) totaled $18,935.01.

Annual Meeting Revenue/Expenses:

Annual Meeting revenue (registration, courses, exhibitor space, dues, BASF award, industry social and sustaining dues) totaled $93,565.00.

Annual meeting expenses (hotel + AV, joint mtg profit distribution, PayPal fees, banner, speakers, easels, awards) totaled $76,792.46.

Education Fund Contributions

Contributions received for the Education Fund totaled $400.00 for the 2016 fiscal year.

NEWSS FISCAL STATEMENT

NOVEMBER 1, 2015- OCTOBER 31, 2016

Income Received	Amount
Joint Meeting Total Registration	$68,945.00
Joint Meeting Courses	$1,920.00
Joint Meeting Exhibitor Space	$1,000.00
Member Dues Only	$150.00
BASF awards	$300.00
Industry Social (4 Sponsors at $3000.00)	$12,000.00
Interest earned on all accounts	$1,043.59
Sustaining Membership: 20 to date	$9,250.00
Sustaining Designated for Weed Contest	$5,000.00
Endowment Contributions	$400.00
Additional Weed Contest Contributions	$2,025.00
Total NEWSS Income Nov 1, 2015- October 31, 2016	**$102,033.59**

Expenses Paid	Amount
PayPal Fees	$1,996.75
Joint Meeting Hotel + AV Cost	$49,445.54
Joint Mtg Profit Distribtuion	$15,879.50
NEPPSC banner	$211.03
Ben Franklin performance	$1,500.00
Alphagraphics - Mtg Programs	$1,373.77
NEPPSC Dinner	$728.23
NEPPSC Registration Desk Assistance	$665.00
Lee Van Wychen NEPPSC Travel Costs	$679.66
Tom Mueller - Quiz bowl host airfare	$436.56
Steve Gylling - NEPPSC honorarium	$500.00
Edzard van Santen - workshop/travel costs	$319.00
Sheraton employee gift cards	$209.90
Shipment of easels	$991.06
Meeting Poster boards	$490.88
EC Dishonesty Bond Insurance Renewal	$359.00
Apex Web Studio - WSSAAbstracts.com - 2015 Mtg	$300.00

Apex Web Studio - WSSAAbstracts.com - 2016 Mtg	$300.00
Webiste "A Small Orange" domain fee for NEWSS org.	$200.00
Administrative	$522.24
WSSA Rep Travel to Summer Board Meeting	$1,000.00
Annual meeting Membership Awards - Etch Art	$540.58
Annual Meeting Poster Awards (1st, 2nd and 3rd place)	$350.00
Annual Meeting Student Paper Awards (1st and 2nd place)	$300.00
Annual Meeting Photo Contest (1st, 2nd and 3rd place)	$175.00
Collegiate Weed Contest host fee	$9,753.77
WSSA Dir. Science Policy YEAR 2015 Dues	$2,500.00
WSSA Dir. Science Policy YEAR 2016 Dues	$2,500.00
CAST dues	$1,500.00
Total NEWSS Expenses paid Nov 1, 2015- October 31, 2016	**$95,727.47**

Total Fiscal Year Summary	
Total Income	$102,033.59
Total Expenses	$95,727.47
Net Gain	**$6,306.12**

NEWSS Account Balances October 31, 2016	
Checking	$53,502.86
Savings	$31,844.57
Education Fund	$23,614.39
Total Net Assets	**$108,961.82**

NORTHEASTERN
WEED SCIENCE SOCIETY

2016 NEWSS BUSINESS MEETING

Meeting called to order by Rakesh Chandran at 5:01pm

New members of the society were asked to introduce themselves, new members including faculty and graduate students.

WSSA President Dallas Peterson addressed the NEWSS membership. He appreciated the opportunity to visit NEPPSC. Dallas thanked the society for their support of WSSA and the director of science policy position. He invited all to attend and participate in the 2016 WSSA annual meeting in San Juan.

Rakesh Chandran gave his presidential address "Time for a name change? - provide some history of the NEWSS

- is Weed Science in its infancy?
- are we in tune with reality and public awareness?
- time for action to improve public awareness, role of weeds, benefits and opportunities
- start a conversation – possible name change

Review of 2015 annual business meeting minutes, motion to accept by Joe Neal, second by Greg Armel, motion carries.

Sudeep Matthew gave the necrology report. Members lost this past year include Ralph Hanson, William Mitchell, Edward Higgins and Robin Bellinder. Bellinder will be honored by the development of an endowment fund at Cornell. Endowment will support graduate student travel and research activities in vegetable weed science.

Rakesh Chandran discussed NEPPSC survey. He thanked the EC for their efforts in planning NEPPSC.

Erin Hitchner gave the treasurer report. Discussed money available for 2015 weed contest and that the EC has approved travel support of $1000 for each team. Balance will be moved forward to support the 2016 weed contest. Financial summary – profit $6500, income $46,173.96, expenses $39,677.56, Education Foundation balance $22,496.65

Jim Steffel gave the Audit Committee report and stated the books are in good order.

Randy Prostak gave the CAST report. Randy thanked the society for allowing him the represent them on the CAST board. Randy introduced Dan Kunkel as his replacement.
Dan Kunkel gave the archive report. New information will be archived at end of the 2016 meeting.

Greg Armel discussed the change to the MOP. A motion by Jim Parochetti was brought forward to change who is responsible for NEWSS website. The motion: The responsibility of the society's webpage will be taken away from the membership chair and given to the education chair. Second by Dan Kunkel, motion carries.

Officer change and passing of the gavel. Rakesh Chandran passed the gavel to Shawn Askew. Shawn Askew presents Rakesh Chandran with plaque.

New business

- Resolution Committee – no suggestions for new resolutions
- New members of the EC
  - Carroll Moseley was elected vice-president
  - Kurt Vollmer was elected treasurer-elect
- Resolution committee: Charles Cahoon, Brian McDonnell, Joe Reed
- Nomination committee: appointed: Kathie Kalmowitz, Brian McDonnell
  elected: Carrie Mansue, Alena Castella

2016 Weed Contest will be hosted by Virginia Tech and be held on July 25 and 26 in Blacksburg, VA

Presentation of the 2016 NEWSS Executive Committee

Adjourn

Author Index

Flessner, Michael L.
Frett, John
Ghantous, Katherine M.
Gill, Stanton A.
Giurleo, Daniel
Gohil, Hemant
Gover, Arthur
Gover, Arthur E.
Grabosky, Grabosky C.
Grande, John
Growe, Anthony M.
Hall, Stacie
Haring, Steven C.
Harlow, Christopher D.
Hashemi, Masoud
Heckman, Joseph
Henry, Gerald M.
Hines, Tommy
Hitchner, Erin
Hlubik, David
Holloway, James
Holshouser, David
Howard, Stott
Hunter, Mitch C.
Isbell, Sarah A.
Jemison, John
Jeranyama, Peter G.
Johnson, Gordon C.
Johnson, Quintin
Karsten, Heather D.
Kaye, Jason P.
Kikkert, Julie R.
Kirt, Austin C.
Klick, Suzanne
Kline, Wes
Klodd, Annie
Kreis, Rachel
Kunkel, Brian A.
Kunkel, Daniel
Law, Eugene P.
Li, Yuan

Liebert, Jeff
Limmer, Matt
Lingenfelter, Dwight
Losey, John
Loudermilk, Erica M.
Loughner, Daniel D.
Loy, James B.
Loyd, Bruce M.
MacDonald, Gregory E.
Maimone, Lisa
Mansue, Carrie
Marble, Chris
Martin, Janel L.
Marvin, Jeff
Mathers, Hannah M.
Mayberry, Alex
Mazzone, Tosh R.
McCahill, Ian M.
McDonnell, Brian F.
Melmaiee, Kalpalatha
Mery, Augusta
Miller, Doug
Mirsky, Steven B.
Molnar, Thomas J.
Morris, Andrew H.
Morris, Scott H.
Mortensen, David A.
Moses, Adrian
Muehlbauer, Megan
Murphy, James A.
Myer, Heidi
Myer, Heidi K.
Neal, Joseph
Northrup, Kristy A.
Ntarlagiannis, Dimitrios
Odero, Dennis
Palmer, Cristi
Park, Bradley S.
Pearson, Brian J.
Peer, Wendy A.
Penn, Chad

Perez, Hector E.
Peterson, Vanelle F.
Pittman, Kara B.
Pittman, Kara B.
Polanin, Nick
Porter, Don
Pyne, Robert
Qin, Zhixuan
Rana, Sandeep
Randhawa, Ranjeet S.
Reasor, Eric H.
Reese, Rachel
Reiners, Stephen
Reinhardt, Theresa A.
Reiss, Emily
Richtmyer III, R J.
Roberts, Dana
Robinson, Judith
Roth, Greg W.
Ryan, Matthew
Saha, Debalina
Saha, Debalina
Sanders, John T.
Sandler, Hilary
Scheckel, Kirk
Schrage, Brandon
Schroeder, Mitchell T.
Schulz, Burkhard
Schuster, Charles F.
Sciarappa, William J.
Scott, Barb
Scott, Barbara
Segobye, Kabelo
Senesac, Andrew
Seyfferth, Angelia
Seyfferth, Angelia
Seyfferth, Angelia L.
Shober, Amy
Shober, Amy L.
Simon, James
Slater, Lee

Smith, Larissa
stalter, richard
Starcher, Stephen
Stefani Fae, Giovani
Stewart, Cody
Stiles, Warren C.
Taylor, Matthew
Taylor, Wilbur
Telenko, Darcy E.
Thomason, Wade E.
Tooker, John
Trigiano, Robert N.
Trower, Tim
Tucker, Kevin A.
Ujcich, Jillian
Vail, Gordon
VanGessel, Mark
Vargas, Jose J.
Vea, Ely
Vietor, Donald M.
Voight, Delbert
Vollmer, Kurt M.
Wallace, Barbara
Wallace, John M.
Ward, Daniel
Weller, Stephen C.
Whitehead, Ross D.
Wyenandt, Andy
Yarborough, David E.
York, Alan C.
Youngerman, Connor Z.

NORTHEASTERN
WEED SCIENCE SOCIETY

Northeastern Weed Science Society (NEWSS)
P.O. Box 25
Woodstown, NJ 08098
USA

ISBN 978-1-5108-4928-0

9 781510 849280

CURRAN ASSOCIATES INC.
proceedings
.com

28th International Carbohydrate Symposium (ICS 2016)

New Orleans, Louisiana, USA
17 - 21 July 2016

I.C.S.2016
XXVIII International Carbohydrate Symposium
July 17-22, 2016 • New Orleans, Louisiana, USA